Applied Control Theory for Embedded Systems

Applied Control Theory for Embedded Systems

by Tim Wescott

ELSEVIER

AMSTERDAM • BOSTON • HEIDELBERG • LONDON
NEW YORK • OXFORD • PARIS • SAN DIEGO
SAN FRANCISCO • SINGAPORE • SYDNEY • TOKYO

Newnes is an imprint of Elsevier

Newnes

Newnes is an imprint of Elsevier
30 Corporate Drive, Suite 400, Burlington, MA 01803, USA
Linacre House, Jordan Hill, Oxford OX2 8DP, UK

Library of Congress Cataloging-in-Publication Data

Wescott, Tim.
Applied control theory for embedded systems / by Tim Wescott.
p. cm. -- (Embedded technology series)
ISBN-13: 978-0-7506-7839-1 (pbk. : alk. paper)
ISBN-10: 0-7506-7839-9 (pbk. : alk. paper) 1. Embedded computer systems--Design and construction. 2. Digital control systems--Design and construction. I. Title. II. Series.
TK7895.E42W47 2006
629.8'9--dc22

2006002692

British Library Cataloguing-in-Publication Data
A catalogue record for this book is available from the British Library.

ISBN-13: 978-0-7506-7839-1
ISBN-10: 0-7506-7839-9

For information on all Newnes publications
visit our Web site at www.books.elsevier.com

06 07 08 09 10 10 9 8 7 6 5 4 3 2 1

Printed in the United States of America

For all my teachers

Contents

Preface

Microprocessors are getting smaller, cheaper and faster. Every day, it is easier to embed more functionality into a smaller space. Embedded processors have become pervasive, and as time goes on, more and more functions that were once implemented with analog circuitry or with electromechanical assemblies are being realized with microcontrollers, ADCs and DACs. Many of these assemblies that are being supplanted by the microprocessor are controlling dynamic processes, which is a good thing, because the microprocessor coupled with the right software is often the superior device.

The worm in the apple is that while many microprocessor based controllers are replacing electromechanical or analog electronic controllers, or are being built new for things that could never be practically controlled, most engineers who are skilled at embedded system design are not acquainted with designing control systems, or have fragmentary, ad-hoc knowledge that falls short of that required to do the job at hand.

This is a book about analyzing and understanding embedded control systems. It is written for the practicing embedded system engineer who is faced with a need to design automatic control systems with embedded hardware, to do so quickly and to produce robust, reliable products that perform well and generate profit for the companies that build them.

What's on the CD-ROM?

The CD-ROM for this book contains two sets of code, and some free analysis software.

The analysis software is SciLab, from the Scilab Consortium (Scilab is a trademark of INRIA). It was used to generate most of the graphs that you see in this book.

The first set of code that you will find on the CD-ROM are the SciLab scripts that were used to generate the graphs. These scripts will provide ready examples should you choose to use SciLab for your analysis.

The second set of code that you will find on the CD-ROM is the C and assembly code that is presented in Chapters 9 and 10. This code consists of many building-blocks for embedded controllers, as well as a swept-sine frequency response measurement package.

1

The Basics

1.1 Control Systems

Control systems are all around us. Any system that does our will with a minimum amount of effort on our part is a control system. Room lighting is a control system—you flip the switch, and the lights go on or off. The steering of a car is a control system—you turn the steering wheel, and the car turns. Home thermostats, traffic lights, microwave oven timers—all of these are control systems.

Automatic control systems are control systems that do some of the thinking for us. If we can set a system to perform some task at a high level of abstraction and have it be responsible for correctly performing the detailed sub-tasks to reach that goal, then that system is an automatic control system. Automatic control systems range from the highly intelligent and complex to the absurdly simple. Everyday examples of automatic control systems include flush toilets, room thermostats, washing machines and the electric power grid. Each one of these systems automatically performs its task without direct operator intervention: the flush toilet runs a measured amount of water into the bowl, then fills its tank to the correct level and stops, a room thermostat turns heat or cooling on and off to keep the room temperature constant, a washing machine performs a specific sequence of filling, agitating, spinning and draining, and the power grid delivers power at a constant voltage and frequency in spite of varying loads.

Closed loop control systems will receive most of our attention in this book. A closed loop control system is one that measures its own output to determine what drive to apply to itself. Closed loop control has some powerful advantages, but it also presents some profound difficulties to the control system designer. With careful design, the advantages can be made to outweigh the difficulties, which is why closed loop control is a popular method of solving engineering problems.

1.2 Anatomy of a Control System

A control system is composed of a number of parts. Throughout the rest of this book I will use a consistent terminology for these parts of control systems that follows the normal terms used by control engineers in English-speaking countries.

Figure 1.1 shows the parts of a closed-loop embedded control system. The software generates commands that are translated into an analog signal by the digital-to-analog converter (DAC). This low-power signal is amplified and applied to the plant. The plant contains an actuator to convert the electrical drive from the amplifier into some useful action in the plant. The plant output is connected to a sensor whose output is applied to an analog-to-digital converter (ADC). The ADC value is read by the control software, which uses this information and the command signal to determine the next drive command to the DAC.

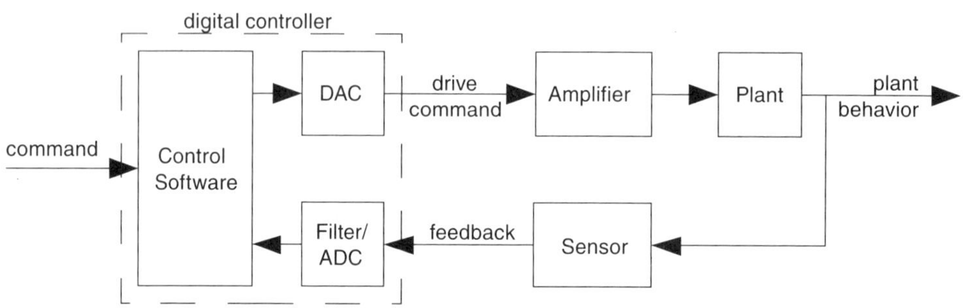

Figure 1.1 A generic control system

Plant

In control-system language, the "plant" is the thing that is being controlled. This name probably originated from the term "steam plant" or "manufacturing plant" as one of the earliest uses of formal control theory was in the regulation of steam plants driving factories. Later on, formal control theory was applied to a variety of other fields, but the term "plant" has stuck.

There is no hard and fast rule for saying exactly which part of your system is the "plant." Depending on the point that you are trying to get across, the word "plant" could be so inclusive to encompass everything in the system except for the core piece of software that implements the controller.

At the very least, however, your control system will contain some item whose behavior you wish to control. This could be as ephemeral as the level of data in a buffer, or as concrete as a thousand-ton passenger train hurtling along at a hundred miles per hour. No matter what it is, this item is, in the end, your "plant." Any more inclusive definition of your plant should include this real plant, and you should not lose sight of what it was that you were trying to get your plant to do in the first place.

It is a good idea, then, when you are describing a control system to state clearly what you mean when you say "plant." Similarly, when you are analyzing a control system, it is a good idea to be flexible about what the term "plant" can mean, and when you are reading a description of a control system you should understand what the writer means when he says "plant."

Controller

Simply enough, a controller is the thing that controls the plant. As with the term "plant," this term can vary in its exclusiveness from the actual core piece of software that implements the controller to everything but the physical hardware that's being affected. As with the plant, it is necessary to be clear in one's definitions, and careful in one's reading of the term.

Through the history of control systems, controllers have been implemented in a number of technologies. The earliest controllers were purely mechanical devices. Even before people thought to apply mathematics to the study of automatic control systems, simple mechanical regulators were being built. In industrial settings, pneumatic controllers were popular for quite a while, because the pneumatic signal from the controller could easily drive a pneumatic actuator. With time, analog electronics came to be used as controllers. Today most new controller designs—even those that masquerade as analog—are digital controllers whose implementation is not much different from Figure 1.1.

In spite of the prevalence of the digital controller, one should not discount other forms. Some systems can benefit immensely from having a mechanical or purely electronic controller to act as a backup to a digital one, or to provide a tight inner loop around which a control loop with a digital controller is placed. It is good to be aware of the places where feedback control already exists, and to know when this can be exploited, or when it must be worked around.

Actuator

At its most exclusive, a controller is just an algorithm running on a microprocessor, reading numbers from a few input ports and writing other numbers to some output ports. At this level a controller does nothing, and can do nothing, to affect the outside world. To perform this task a control system needs an actuator.

An actuator is a transducer that takes a command from the controller and converts it to a useful form of energy to drive the plant. In Figure 1.1 the DAC is what the controller sees as its "actuator." More usually, one thinks of actuators as being some form of motor or other device that takes an already-amplified electrical signal and turns it into useful work.

Actuators can take many forms. Some of the most imaginative and creative work that is done in real-world control system design is often found in actuator selection Also, alas, some of the most mundane and boring is found here as well. Actuators can range from the simple, such as a resistive element in a heating system, to the sublime, such as a Pockels cell in a laser system, which requires several hundred volts to change the polarization of incoming light so that a laser's intensity (or some other characteristic) can be affected.

Sensor

The complement to an actuator is a sensor. Without an actuator, a controller is crippled. Without a sensor it is blind.

A sensor is the transducer that measures some output of the plant and converts it into a form that can be read, directly or indirectly, by the controller. Sensors will usually be analog, where some physical quantity will be translated into a voltage to be read by a DAC. Sometimes, however, a sensor may be able to more or less directly convert a physical quantity into a digital signal; for example, a shaft encoder or a can-counter on a production line.

1.3 Closed Loop Control

Closed loop control systems are a mixed blessing, particularly for the engineer who has to design or maintain one. Closed-loop control has many advantages, but for every advantage there is a drawback, and threaded through any control system problem are the limitations imposed by the basic physics of the plant, sensors, actuators and controller.

If you can accurately measure a plant's output with a sensor, and if you can reliably induce changes in the plant's output to any degree necessary, then you can build a controller that will position the plant with a degree of accuracy that approaches that of the sensor—even if the plant, by itself, cannot be positioned reliably at all.

With your accurate sensor and well-behaving plant, your controller can respond to plant errors that arise from external disturbances that perturb your plant's output. Even if the plant is so sensitive to disturbances that it is useless without closed-loop control, if it is responsive to control you can build a system that, as a whole, can operate correctly even with continuous disturbances.

If your plant can respond strongly to drive, but is, by itself, sluggish, a good feedback control system will speed up the plant's response. This fast plant response can be used to follow a command signal better, or simply to reject outside disturbances better.

In addition to being able to make an accurate plant out of an inaccurate one, given a good sensor you can turn this around: with a plant that responds predictably both to drive and to external stimuli, but with a sensor of limited range, you can build a wide-ranging, accurate sensor for physical phenomena. Specifically, force-balancing accelerometers and rate integrating gyros perform this task by keeping a proof weight (or spinning wheel) centered within the frame of the device with an accurate force (or torque) driven by closed-loop control.

Every silver lining has a cloud, however, and for all its advantages closed-loop control systems have some disadvantages and some limitations.

The most dramatic disadvantage to closed loop control is instability. You want to have a system that goes where it is supposed to and then stays put. With closed-loop control, however, you can take a perfectly stable plant and a perfectly stable controller and put them together into a shaking, snarling beast of a closed-loop system that will loop its own errors back on itself and oscillate around its desired setpoint in ways severe enough to harm actuators, mechanisms, or even the surrounding scenery.

If you take care with your control system design you can avoid instability—but such care takes development time and money. In addition, when you add a controller to a system you add direct recurring costs to purchase the controller's parts, assemble them, and make sure that their design stays current.

A final 'problem' in this list is that control systems are not magic. Real systems have real limitations, and nothing—not even an automatic control system—can overcome the laws of physics. Limitations in the strength of your actuators, in the response speed of every part of the system, in your plant's tendency to allow itself to be driven nicely, and in your sensor's ability to deliver noise-free measurements will all limit the ultimate performance of your system.

For all these problems and limitations, closed-loop control systems are an everyday part of life. Why? Because for many situations the advantages of closed-loop control systems far outweigh their disadvantages. Where a closed loop system may go unstable an uncontrolled plant may not work at all. Where a closed-loop control system is sensitive to sensor problems, its open-loop counterpart would be buried under plant variations. The biggest reason to use a closed-loop control system is because even with the expense and complexity of a controller, you can often do the overall job with much less expense, and much higher quality, by using closed-loop control.

1.4 Controllers

Executives

Executive controllers are the kind of controllers that most of us are familiar with designing. In fact, the usual meaning of the 'controller' in 'microcontroller' is an executive controller.

An executive controller functions much as an executive in a company. An executive controller is concerned with making the right thing happen at the right time, and is content with issuing certain orders at certain times. If an executive controller *does* get feedback on the things that it is controlling, it is feedback that is at an abstract level; "motor two is broken" or "process four is complete." An executive controller issues commands at a similarly abstract level; the deepest it may go is to command something to go to a certain value, while expecting that its command will be followed.

This book is, by and large, not concerned with executive controllers. It is not concerned with executive controllers for three reasons: first, if you are reading this book you either already know how to design executive controllers or will soon meet people who will teach you; second, the design of executive controllers can be done quite effectively in an ad-hoc, heuristic manner; and finally, a precise mathematical treatment of executive controller design would occupy a more than a few feet of shelf space, if it could be formulated at all.

Open-Loop Controllers

If a controller just sets some level of drive to a plant without paying attention to the plant's behavior then the control is "open-loop." Executive controllers are often open-loop, or essentially so. Any time that the control to the plant is set without using knowledge of the plant state, the control is open loop.

Regulators

A regulator is a controller that monitors the plant output and varies the command to the plant to hold the plant output at some desired level, or *set point*. In general, a regulator's set point does not change very often, if at all. The only important dynamic behavior of a regulator system is how rapidly and reliably it reaches the set point when the system starts up, and how well it responds to disturbances on its output.

Because a regulator drives the plant with information (feedback) from the plant, the system as a whole forms a loop; the common name for a control system with feedback is a *control loop*.

Servo Systems

A servo system is like a regulator in that it works in closed-loop mode, monitoring the plant output and varying the command to the plant. The difference is that the job of a servo system is to cause the plant output to follow the input command as closely as possible, not just hold the output steady to some set point. A servo system not only needs to hold its output accurately, but it must follow the commanded input faithfully and quickly.

Hierarchical Controllers

It is rare to find a control system that doesn't have some hierarchy of control. A system that uses a control loop as a plant for a larger control loop is not uncommon, while it is quite common indeed to see an executive controller driving a regulator or a servo system.

1.5 About This Book

This book can be divided into roughly two parts. Chapters 1 through 5 are almost purely control theory. Starting with Chapter 6, the focus of the book starts to slide into the practical, arriving there fully in Chapters 9 and 10.

Chapter 1 contains introductory material. It defines a control system, and outlines some of the themes that will be presented later in the book.

Chapter 2 covers the z transform. The z transform is the mathematical foundation for discrete-time control, and this chapter forms the mathematical foundation for the rest of the book. Chapter 2 introduces the z transform, shows how it can be used to solve basic problems in control theory if one assumes a linear system, and shows ways that it can be used in a way that is, if not painless, then at least fairly direct and easily.

Chapter 3 covers performance criteria for control systems. This chapter presents the measures of performance that are most commonly used (and useful) for describing a control system's performance. It shows how these performance criteria are arrived at, how they can be measured, and what they imply in terms of the system's characteristics as described using the z transform.

Chapter 4 covers how one describes a control system using block diagrams. It shows the block diagramming language as it is used by control systems engineers, and it shows how a properly-constructed block diagram can be used to analyze a system's behavior in a clear and concise manner.

Chapter 5 is about analyzing control systems. It shows how to use a control system description to arrive at results that go beyond specific predictions for individual systems, allowing the control system designer to predict how a particular control system design will perform in the face of changing system characteristics whether these changes arise from aging, different uses of the system, or manufacturing variation.

Chapter 6 covers the design of control systems. It shows a number of commonly used elements for controllers, and how these controller elements affect the aspects of system behavior specified in Chapter 3 and analyzed in Chapter 5.

Chapter 7 is about sampling. In Chapter 2, the assumption is made that sampled data is available, and this assumption is carried through Chapter 6. Chapter 7 shows how to go about taking a real-world plant that exists in continuous time, and converting its behavioral description into a discrete-time model that can be used with the lessons learned in Chapters 1 through 6.

Chapter 8 covers nonlinear control. Most control system design is done assuming that a plant (and controller) are linear, yet the world is a nonlinear place. Chapter 8 shows how to resolve the conflict between the easy-to-use linear control system tools with the nitty-gritty real world of nonlinearities.

Chapter 9 shows one method of measuring a plant's behavior. One cannot perform rational, directed control system design without knowing how a plant behaves. Unfortunately it is often not practical to model a plant from first principles. Chapter 9 shows the frequency response method of measuring a plant's behavior, in a way that can be done practically with the very control system fabric that one is developing.

Chapter 10 caps off the book with a description of the methods used to translate control systems designs into software. It discusses some of the pitfalls of software design, primarily the peculiarities of working in an environment with fixed numerical precision. It includes some recommended controller architectures that have worked well for the writer in the past.

2

Z Transforms

Control system design is intimately involved in dynamic systems—after all, the point of a control system design is to take a system and impose your will on its dynamics, so that it will do what you need it to do. As such, any exercise in control system design boils down to an exercise in getting useful information out of differential or difference equations. So the end product of control system design is a set of information that either contains solutions to difference (or differential) equations, or contains information about these solutions, or both.

When working with sampled-time control problems, it is the difference equation that becomes the most important. The difference equations in full-blown system models that account for every nuance of a system's behavior are generally impossible to formulate, let alone to solve. We can, however, make some important simplifications to our system models that will yield equations that are easy to solve or (often better) to extract general information from.

Most of this book uses the z transform either directly or indirectly to analyze and design control systems. The z transform is a powerful method for solving linear, shift invariant difference equations. For the many systems that can be adequately modeled by such difference equations, the z transform is an invaluable tool for control system design.

This chapter introduces the z transform and presents its power and its limitations. Subsequent chapters will build on the information in this chapter to show how to use the z transform to analyze, design and implement embedded digital control systems.

In every chapter of this book I have tried to present the material in as understandable a way as possible, and given as many immediately useful results as I can. I will fall the shortest of this goal in this chapter. The mathematics are as involved as any you'll find anywhere in the book, and there are only a few results that are directly applicable. The reason that this chapter is here, in the book and in this position, is because much of the material in the following four chapters depends on the z transform.

2.1 Signals and Systems

This book will talk extensively about signals and systems, so it is essential that we define these terms.

The term "signal" as used in this book, refers to a stream of information (either in continuous time or in discrete samples) that evolves over time. Signals can be present as the value of a register in hardware, a variable in software, as a voltage on a circuit node, as a pressure, or temperature, or any other physical variable. Anything that has a numerical value, the meaning of which is germane to the problem at hand, can be viewed as a signal. The important aspects in a signal are that its value means something to us, its evolution in time is affected by other signals, and that it can affect other signals in turn.

When dealing with control systems, one should draw a distinction between a real physical phenomenon and the signal that represents it in a control system model. For example, the velocity and position of a log in a sawmill have values which can be regarded as signals. These signals are interesting mathematical abstractions and can do no damage to anyone at all. On the other hand, if you're in a sawmill to debug an embedded controller in a machine and there is a log hurtling toward you at high speed, the actual physical phenomena of its position and velocity can have a great impact on your well being.

When doing control system design one must work with the signals, but it is good to remember that what really matters in the outside world is the thing itself. Thus if you are talking about how a plant responds to control you may talk about a "drive signal"—but it may be more appropriate to talk about the "drive" itself, particularly if you are discussing some aspect of the real system (such as heating, or parts flying around the room) that is not covered in the control system model.

In this book a signal will be denoted as a function of time (or sample) or as a variable. Continuous-time signals will be denoted as a function of time, $x(t)$. Discrete-time signals will be denoted as a function of their sample number, $x(n)$ or as a subscripted variable x_n. In both the continuous time and discrete time cases, a signal may be indicated without the time dependence: once I establish that $x(t)$ is a signal in continuous time, for example, I may continue to talk about it as x, leaving the reader to remember that it's a signal.

In this book, the word "system" will have two meanings: one is the more popular meaning of a conglomeration of parts that act together to achieve some goal (such as a "control system"); the other meaning of the word "system" is a mathematical construct that acts on signals. In this second meaning, a system takes one or more signals as inputs and generates one or more signals as outputs. The value of the output signal(s) at any one time depend on the history of the input signal(s) over time.

If you have a signal u and a system h, running u through h will produce a new signal:

$$x(t) = h\left(u(t), t\right). \tag{2.1}$$

If one is being strict about one's definitions, there are only three things that define x: the behavior of the system h, the time history of the signal u, and the time t. No other thing should matter in this definition; if it does then the description is inadequate. Note that when you are representing some real-world phenomenon as a system, there is nothing wrong with purposely leaving out some effect—this is an essential part of simplifying the world so we can analyze it. Just as long as the effect you are neglecting doesn't matter you'll be fine.

Here are some example systems:

A power meter would measure some physical quantities such as voltage and current, multiply them, and generate a time average of the result:

$$p_{measured}(t) = \frac{1}{T}\int_{t-T}^{t} v(\tau)i(\tau)d\tau. \tag{2.2}$$

An integrator in an embedded controller would take its input signal, multiply it by some constant, and sum the result over all time from system startup:[1]

$$y(n) = \sum_{k=0}^{n} k_i x(k). \tag{2.3}$$

A multiplier might take two signals and return their product:

$$y(t) = x_1(t) x_2(t). \tag{2.4}$$

Linearity

The z transform only works on systems that are linear. A linear system is one that obeys the property of superposition, i.e., for any two signals x and y and for any two constants a and b, system h is linear if and only if

$$h\left(ax(t) + by(t)\right) = ah\left(x(t)\right) + bh\left(y(t)\right). \tag{2.5}$$

For example, the system $h(x) = 25x$ is a linear system because it obeys (2.5); the system $h(x) = x^2$ is not. Another way of viewing this is that a system that can be represented by a linear difference or differential equation is linear, while a system that must be represented by a nonlinear difference or differential equation is not linear. Most writers prefer the definition in (2.5) because it is more general, more definitive, and because it is a test that you can perform on a real system[2] without having to have a detailed mathematical model of the system.

The property described in (2.5) is called *superposition*, and it makes the analysis of linear systems much easier than that of nonlinear systems. Using superposition you can separate the effects of various factors on a linear system and analyze them in isolation, then combine their effects at the end to find a solution. This gives you a tremendous ability to reduce the amount of analysis that you must do, and allows for finding general solutions to problems.

One result of superposition is that the behavior of a linear system is global. Global behavior means that the basic response of the system to external stimuli

---

1 In the real world, of course, the integrator wouldn't store a vector of all past values of x; it would just add k_i times x to y at each iteration of the control loop.

2 If you do perform this test, you can only prove linearity over the range of inputs for which you have done the test. This is, however, often quite sufficient.

and to its own internal states will always be the same, no matter how many different effects are present at the same time.

Shift (In) Variance

In addition to only working with linear systems, the z transform also requires the system to be shift invariant. Indeed, we will only discuss the analysis of shift invariant systems in this chapter. A shift invariant system is one that doesn't change its behavior at different time; in other words, given an input signal x_k and a system h, if h generates

$$y_k = h(x_k), \tag{2.6}$$

then h is shift invariant if and only if

$$h(x_{k-\kappa}) = y_{k-\kappa}\,. \tag{2.7}$$

For example, the system $h(x_k) = x_k^2$ is shift invariant, but the system $h(x_k) = \sin(\theta k)x_k$ is not.

Note that the properties of shift invariance and linearity are independent: a system can be nonlinear and shift varying, it can be either linear or shift invariant, or it can be both linear and shift invariant.

Continuous-time systems are defined in a similar manner, being either time invariant or not. Time invariance is a stricter condition than shift invariance: if you perform analysis or measurements on the continuous-time portion of a system with an embedded controller you will find that it is time varying because of the action of the embedded controller. If a system has time varying behavior that is periodic and synchronized to the sample rate then it will be shift invariant but not time invariant.

2.2 Difference Equations

Time in the real world is a continuous process. Thus, when one sets about to model a physical system in the real world one naturally uses differential equations because they accurately model continuous processes.

For a digital controller, the situation is different. Digital systems divide time into discrete processor cycles, and it takes many such cycles to execute a program. While we can always break our intervals down into smaller and smaller time quanta, doing so requires that we spend more and more processor cycles on each second of real time. Since processor cycles cost money, attempting to use digital hardware to implement true differential equations would lead to infinitely large and expensive hardware, which is hardly desirable. So when we deal with control problems in the discrete-time world we use discrete-time equations.

The discrete-time equivalent to the differential equation is the difference equation. Where the differential equation describes a system based on the speed at which each of the variables is changing, the difference equation describes a system based on what's going to happen at the next tick of the clock. Because of this, one can design one's difference equations to execute in a known amount of time.

Even a differential equation solver on a digital computer uses difference equations, albeit much more sophisticated than we'll discuss here (and much less certain in the execution-time department).

(2.8) shows a difference equation that you might see in practice. It says that the current value of the output signal x is equal to the constant a_1 times the previous value of x plus the constant a_2 plus the next earlier value of x, plus constants b_1 and b_2 times the two previous values of the input signal u:

$$x_k = a_1 x_{k-1} + a_0 x_{n-2} + b_1 u_{n-1} + b_0 u_{n-2} \,. \tag{2.8}$$

In general the difference equations that we will be treating here are of the form

$$x_n = \sum_{1}^{K} a_{K-k} x_{n-k} + \sum_{0}^{K} b_{K-k} u_{n-k} \,. \tag{2.9}$$

In other words, the current value of the output will be equal to a weighted sum of previous values of the output, plus a weighted sum of previous (and perhaps current) values of the input.

Solving

Just as linear differential equations can be solved directly, so can one solve linear difference equations directly. I won't dwell on this here, because the z transform provides a much better and more systematic approach to solving complex linear shift invariant difference equations, but I will give an outline of how it is done.

Example 2.1 Solving Difference Equations

A system is described by the difference equation

$$x_n = 1.85x_{n-1} - 0.855x_{n-2} + 0.005u_n, \tag{2.10}$$

where the signal u is 0 for all $n < 0$, 1 for $n \geq 0$, and where $x_{-2} = x_{-1} = 0$. Find the values of x_n for all $n \geq 0$.

As with differential equations, the solution to (2.10) has both homogeneous and nonhomogeneous solutions. Bringing all of the x terms to the left side results in

$$x_n - 1.85x_{n-1} + 0.855x_{n-2} = 0.005u_n. \tag{2.11}$$

As with a differential equation, we can find the homogeneous solution with an auxiliary polynomial; in this case the auxiliary polynomial is

$$q^2 - 1.85\text{q} + 0.855 \tag{2.12}$$

which has roots at $q = 0.9$ and $q = 0.95$. This translates to a homogeneous solution to (2.11) of

$$x_h = a_1(0.9)^n + a_2(0.95)^n. \tag{2.13}$$

In typical differential-equation fashion, the nonhomogeneous solution of (2.11) is a constant, so the overall solution is

$$x_n = a_0 + a_1(0.9)^n + a_2(0.95)^n. \tag{2.14}$$

Inserting (2.14) into (2.11) and simplifying we get

$$0.005a_0 = 0.005, \tag{2.15}$$

from which we can easily conclude that $\text{a}_0 = 1$.

2.3 The Z Transform

One can use the Laplace transform to solve linear time invariant differential equations, and to deal with many common feedback control problems using continuous-time control. With a sampled-time system one deals with linear shift invariant difference equations, and the tool for analysis is the z transform.

By definition, the z transform takes an expression for a signal x_k which is dependent on the time variable k and transforms it into an expression $X(z)$ that depends only on the variable z. Note that the signal hasn't changed; its description will look totally different, but the z transform preserves all information about the signal. That's the transform part: the problem is transformed from one in the sampled time domain k, and put it into the z domain.[3] The z transform of x is denoted as Z{x}. In this book we'll use the single-ended z transform which is defined as:

$$X(z) = \mathbf{Z}\{x\} = \sum_{k=0}^{\infty} x_k z^{-k}. \tag{2.16}$$

Because we have defined the transform as starting at $k = 0$ the signal must be restricted to $x_k = 0$ for all $k < 0$. In practical control-systems problems this restriction causes no problems, and saves us from having to worry about what's been happening to our signals since before the beginning of time! As an example of the z transform, if you have a signal

$$x_k = \begin{cases} 0 & k < 0 \\ a^k & k \geq 0 \end{cases} \tag{2.17}$$

you can plug it into (2.16) and get:

$$X(z) = \sum_{k=0}^{\infty} a^k z^{-k} = \frac{z}{z-a}. \tag{2.18}$$

Obviously (2.16) can be used to find the z transform for any arbitrary signal—you just may not be able to get it into a nice closed form as was done in (2.18), which is tedious, however. In practice one is interested in using signals

3 Later in this chapter we will see how the z domain can be viewed in terms of frequency. For now just take it as a mathematical convenience.

to investigate the general behavior of a system; in this case one picks signals that have easy closed-form transforms. Table 2.1 lists a number of time-domain signals and their z transforms.

2.4 The Inverse Z Transform

Unlike the z transform given in (2.16), there is no useful, mathematically tidy and general expression that will get you from the z domain back to the time domain. There are, however, two very useful techniques that will always work if the signal in question is expressed as a ratio of polynomials in z. One uses the fact that control systems tend to generate responses which are easy to categorize and identify; since there is a one to one relationship between a time-domain signal and its z transform, the inverses can be found by inspection after a little work. The other method is a "brute force" one of synthetic polynomial division, which will always yield an answer when the z transform is in the form of a ratio of polynomials.

By Identification

When one solves problems in control systems using the z transform, one invariably gets results that are expressed as ratios of polynomials in z. Indeed, (2.18) is an example of a z transform that results in a small polynomial ratio. By using partial fraction expansion, these polynomial ratios can be broken down into terms that can be found in Table 2.1; because there is a one-to-one correspondence between a signal and its z transform, one can identify the time-domain signals from their z transforms to construct the time-domain signal.

Partial fraction expansion is based on the basic grade-school rule of adding fractions

$$\frac{x}{a} + \frac{y}{b} = \frac{xb + ya}{ab}, \tag{2.19}$$

applied to polynomials. If a and b in (2.19) are real first-order polynomials and x and y are real numbers then the numerator of the resulting fraction is a real number or a first-order polynomial, and the denominator is a second-order polynomial. In fact, one can sum N distinct first-order polynomial ratios together; the result is a polynomial ratio with a numerator of order N-1 or less and a denominator of order N.

Less obviously, if one has a proper[4] polynomial ratio with distinct roots in the denominator, it can always be split into a sum of first-order polynomial ratios (sometimes at the cost of having ratios with complex roots). Furthermore, if we have a polynomial that has repeated roots it can be split into the form

$$\frac{N(z)}{(z-d)^n(\cdots)} = \frac{A_1 z}{z-d} + \frac{A_2 z}{(z-d)^2} + \cdots + \frac{A_n z}{(z-d)^n} + \cdots. \tag{2.20}$$

Example 2.2 Partial Fraction Expansion

A system response can be split into its constituent parts by directly solving (2.19). Take a system response given by

$$X(z) = \frac{0.1z}{z^2 - 1.9z + 0.9}. \tag{2.21}$$

To find the partial-fraction expansion of this system response, first factor the denominator to find that it has roots at $z = 1$ and $z = 0.9$. Then substitute the resulting first-order polynomials into (2.19):

$$X(z) = \frac{Az}{z-1} + \frac{Bz}{z-0.9} = \frac{Az - 0.9A + Bz - B}{(z-1)(z-0.9)}. \tag{2.22}$$

Now the terms can be collected to form a system of two equations in two unknowns, which can be solved for A and B:

$$\begin{aligned} A + B &= 0 \\ -0.9A - B &= 0.1 \end{aligned}. \tag{2.23}$$

Do the math, and you find that $A = 1$ and $B = -1$, so (2.21) resolves to

$$X(z) = \frac{0.1z}{z^2 - 1.9z + 0.9} = \frac{z}{z-1} - \frac{z}{z-0.9}. \tag{2.24}$$

---

[4] A proper polynomial ratio is one where the order of the numerator is less than the order of the denominator.

The method shown in Example 2.2 is a good one, but it gets very cumbersome for large polynomial ratios. This is fine if one is using a computer to do the expansion, but if one should need to do so by hand it can be tedious. At the cost of more difficult mathematics, there is a better way of doing this.

The expression

$$X(z) = \frac{N(z)}{(z-a_1)(z-a_2)(\cdots)(z-a_N)} = A_0 + \frac{A_1 z}{z-a_1} + \frac{A_2 z}{z-a_2} + \cdots + \frac{A_N z}{z-a_N}, \quad (2.25)$$

puts X into a form whose parts match the entries in Table 2.1. As long as we have a way of finding A_0 through A_N then we can find our inverse z transform. To find A_1 through A_N, multiply both sides of the equation by the expression $(z - a_1)/z$ and find its value as z approaches a_1. You can see that

$$\lim_{z\to a_1}\left[\frac{z-a_1}{z}X(z)\right] = \lim_{z\to a_1}\left[\frac{A_0(z-a_1)}{z} + \frac{A_1(z-a_1)}{z-a_1} + \frac{A_2(z-a_1)}{z-a_2} + \cdots\right]. \quad (2.26)$$

But this resolves to zero for every term whose denominator doesn't equal $z - a_1$, so

$$\lim_{z\to a_1}\left[(z-a_1)X(z)\right] = A_1\,. \quad (2.27)$$

This suggests a technique for expanding an arbitrary signal: For each root in the denominator, omit that root and evaluate the resulting expression for z equal to the root—this gives you the coefficient (called the residual) for that root's contribution to the expansion. This leaves the A_0 term: logically you can see that if $N(z)$ has no constant term then

$$X(z)z^{-1} = \frac{N(z)z^{-1}}{(z-a_1)(z-a_2)(\cdots)(z-a_N)} = 0 + \frac{A_1}{z-a_1} + \frac{A_2}{z-a_2} + \cdots + \frac{A_N}{z-a_N} \quad (2.28)$$

and you can safely conclude that $A_0 = 0$. If $N(z)$ *does* have a constant term then A_0 must be found. Appealing once again to grade-school mathematics, we find that

$$\frac{A_1}{z-a_1} + \frac{A_2}{z-a_2} + \cdots + \frac{A_N}{z-a_N} = \frac{(A_1 A_2 \cdots A_N)z^N + \cdots}{(z-a_1)(z-a_2)(\cdots)(z-a_N)} \quad (2.29)$$

and

$$A_0 = \frac{A_0(z-a_1)(z-a_2)(\cdots)(z-a_N)}{(z-a_1)(z-a_2)(\cdots)(z-a_N)} = \frac{A_0 z^N + \cdots}{(z-a_1)(z-a_2)(\cdots)(z-a_N)}. \quad (2.30)$$

We can equate the z^N terms of the left half side with the z_N term of the numerator to get

$$\frac{(A_0 + (A_1 A_2 \cdots A_N))z^N + \cdots}{(z-a_1)(z-a_2)(\cdots)(z-a_N)} = \frac{n_N z^N + \cdots}{(z-a_1)(z-a_2)(\cdots)(z-a_N)}. \quad (2.31)$$

Solving for A_0 gets us

$$A_0 = n_N - (A_1 A_2 \cdots A_N). \quad (2.32)$$

This method of using partial fraction expansion is very nice because it saves you from juggling large polynomials and linear systems of equations as would be necessary in the method presented in Example 2.2. So far, however, this technique is missing an important element: it cannot deal with repeated roots.

In order to get to a polynomial expansion that can find the coefficients of the expansion for repeated roots, and implied by (2.20) we need to augment our method. This is done by multiplying the expression by the multiple root, then taking the derivative with respect to z, as many times as is necessary to extract all of the roots. Taking an example with a second-order root:

$$X(z) = \frac{N(z)}{(z-a_1)(z-a_2)^2} = A_0 + \frac{A_1 z}{z-a_1} + \frac{A_2 z}{(z-a_2)^2} + \frac{A_3 z}{z-a_2}, \quad (2.33)$$

finding the coefficient A_1 works as before. Finding the coefficient A_2 is done by multiplying by the double root:

$$\lim_{z \to a_2} \left[\frac{(z-a_2)^2}{z} X(z) \right] = \frac{(z-a_2)^2 A_0}{z} + \frac{(z-a_2)^2 A_1}{z-a_1} + \frac{(z-a_2)^2 A_2}{(z-a_2)^2} + \frac{(z-a_2)^2 A_3}{z-a_2} = A_2. \quad (2.34)$$

Then

$$\lim_{z\to a_2}\left[\frac{d}{dz}\frac{(z-a_2)^2}{z}X(z)\right]=\lim_{z\to a_2}\frac{d}{dz}\left[\frac{(z-a_2)^2A_0}{z}+\frac{(z-a_2)^2A_1}{z-a_1}+\frac{(z-a_2)^2A_2}{(z-a_2)^2}+\frac{(z-a_2)^2A}{z-a_2}\right.$$

$$\lim_{z\to a_2}\left[\frac{d}{dz}(z-a_2)^2X(z)\right]=\frac{d}{dz}\left[\frac{(z-a_2)^2A_0}{z}+\frac{(z-a_2)^2A_1}{z-a_1}+A_2+(z-a_2)A_3\right] \quad .(2.35)$$

$$\lim_{z\to a_2}\left[\frac{d}{dz}(z-a_2)^2X(z)\right]=A_3$$

This procedure can be extended indefinitely, so for (2.20), finding the coefficient for the mth repeated root out of n repeated roots is a matter of finding

$$A_m=\frac{1}{(n-m)!}\frac{d^{n-m}}{dz^{n-m}}\frac{(z-a)^n}{z}X(z). \quad (2.36)$$

The procedure for finding A_0 remains the same.

x_k	$X(z)$	Comments	
$d(k)$	1	$d(k)=\begin{cases}1 & k=0\\ 0 & k\neq 0\end{cases}$ [5]	(2.37)
$u(k)$	$\frac{z}{z-1}$	$u(k)=\begin{cases}0 & k<0\\ 1 & k\geq 0\end{cases}$ [6]	(2.38)
$ku(k)$	$\frac{z}{(z-1)^2}$	$ku(k)$ is a sampled unit ramp function.	(2.39)
$k^2u(k)$	$\frac{z}{(z-1)^3}$		(2.40)
$k^{n+1}u(k)$	$\lim_{b\to 0}\left[(-1)^n\frac{d^n}{db^n}\frac{z}{z-e^b}\right]$	This is nasty, I haven't found a better one yet.	(2.41)

Table 2.1 Some common z transforms (continued on next page)

[5] $d(n)$ is the z transform equivalent of the Dirac delta function.

[6] $u(n)$ is known as the "unit step function." Note that it *is* defined at $n = 0$.

x_k	$X(z)$	Comments	
$d^k u(k)$	$\dfrac{z}{z-d}$	d must be a constant.	(2.42)
$kd^k u(k)$	$\dfrac{z}{(z-d)^2}$	d must be a constant.	(2.43)
$d^k \cos(\theta k) u(k)$	$\dfrac{z^2-(d\cos\theta)z}{z^2-2d\cos(\theta)z+d^2}$	complex pole, d and θ must both be constants.	(2.44)
$d^k \sin(\theta k) u(k)$	$\dfrac{d\sin(\theta)z}{z^2-2d\cos(\theta)z+d^2}$	complex pole, d and θ must both be constants.	(2.45)

Table 2.1 Some common z transforms (continued from previous page)

By Synthetic Division

The most useful and commonly used property of the z transform is that the z transform of a signal that is delayed by one sample time is the z transform of the original signal times z^{-1}:

$$X(z_{k-1}) = \sum_{k=0}^{\infty} x_{k-1} z^{-k} = \sum_{k=0}^{\infty} x_k z^{-k-1} = z^{-1} \sum_{k=0}^{\infty} x_k z^{-k} = z^{-1} X(z_k). \tag{2.46}$$

In fact, we can use just this property to show how to find an original signal by synthetic division. Synthetic division works like long division: you evaluate the expression to get the leading z term and a remainder, then you repeat again, as necessary until you're bored or until your problem is solved.

Example 2.3 Inverse Z Transform by Synthetic Division

Take the system response from Example 2.2:

$$X(z) = \frac{0.1z}{z^2 - 1.9z + 0.9}. \tag{2.47}$$

To divide this out we just start multiplying it by z^{-1} and keeping the fraction proper:

$$X(z) = z^{-1}\left(\frac{0.1z^2}{z^2 - 1.9z + 0.9}\right) = 0.1z^{-1} + z^{-2}\left(\frac{0.19z^2 - 0.09z}{z^2 - 1.9z + 0.9}\right) + \cdots. \tag{2.48}$$

By carrying this out you get

$$X(z) = 0.1z^{-1} + 0.19z^{-2} + 0.271z^{-3} + 0.6561z^{-4} + \cdots. \tag{2.49}$$

This numerical result is, in fact, the solution to (2.24) that was found in Example 2.2, except that it is all done numerically.

This method of synthetic division will always yield a result without having to go through the pain of a partial fraction expansion, but what you end up with is a numerical result that doesn't give you much insight into the behavior of the system.

2.5 Some Z Transform Properties

This section lists a few of the properties of the z transform. Some of these will simplify the construction or solution of a problem. Some of these will let you extract information about a signal without having to know the exact detail of the signal. Nearly all of them can be derived from (2.16) or (2.46).

Delay

This is such a basic property of the z transform that I've already derived it, in (2.46) in the section on synthetic division. The fact that a delay of one step in the time domain causes a multiplication by z^{-1} in the z domain is one of the fundamental properties that makes the z transform so useful.

Linearity

We have, in fact, already used this property of the z transform to find solutions to the inverse z transform problem using partial fraction expansion. To be linear, the z transform must satisfy (2.5). Combining (2.5) (the definition of linearity) and (2.16) (the definition of the z transform) gives us

$$\sum_{k=-\infty}^{\infty}(ax_k+by_k)z^{-k}=\sum_{k=-\infty}^{\infty}ax_kz^{-k}+\sum_{k=-\infty}^{\infty}by_kz^{-k}=aX(z)+bY(z), \tag{2.50}$$

so the z transform is linear (this follows from the linearity of the summation operation).

The fact that the z transform is linear means that if we have the z transforms of two signals, we can know the z transform of their sum by simple addition. This property will allow us to manipulate our systems' difference equations in the z domain without having to go to the time domain and back, which will greatly simplify system analysis.

First Difference

It is common when designing control systems to use a first difference operation (often referred to as "differentiator" from its continuous-time equivalent) as a means of anticipating the behavior of the plant. If we define a signal y to be the first difference of x, like so:

$$y_k=x_k-x_{k-1}, \tag{2.51}$$

then we can find the z transform easily by using (2.46):

$$Y(z)=X(z)-z^{-1}X(z)=\frac{z-1}{z}X(z) \tag{2.52}$$

Summation (Integration)

Even more common than the first difference, control systems often use the summation of a signal. Almost universally referred to as "integration" for its continuous-time equivalent, this operation can keep track of a system's errors over all time, to make sure that they are driven to zero.

Define the operation to be

$$y_k = \sum_{p=-\infty}^{k} x_p \,. \tag{2.53}$$

Taking the z transform, using (2.16) to get

$$Y(z) = \sum_{k=-\infty}^{\infty} z^{-k} \sum_{p=-\infty}^{k} x_p = \sum_{k=-\infty}^{\infty} \sum_{p=-\infty}^{k} x_p z^{-k} \tag{2.54}$$

now change the variable in the final summation and reverse the order of summation to get

$$Y(z) = \sum_{k=-\infty}^{\infty} \sum_{q=0}^{\infty} x_{k-q} z^{-k} = \sum_{q=0}^{\infty} \sum_{k=-\infty}^{\infty} x_{k-q} z^{-k} \,. \tag{2.55}$$

Now we can see that the second summation is just the z transform of x delayed by q samples, so we can use (2.46) to get

$$Y(z) = \sum_{q=0}^{\infty} X(z) z^{-q} = X(z) \sum_{q=0}^{\infty} z^{-q} = \frac{z}{z-1} X(z) \,. \tag{2.56}$$

Final Value Theorem

Often you will have a z transform of a signal and you will only care about a few aspects of its time-domain behavior. Rather than having to find an exact representation of the signal, you are often only interested in how quickly it settles to some final value, and what that final value is.

The final value theorem states that for any signal x_k that is zero[7] for all $k < 0$, has a finite value as x_k goes to infinity and has a z transform $X(z)$, the value of x_k as k goes to infinity is equal to

$$\lim_{k \to \infty} x_k = \lim_{z \to 1} \left[(z-1) X(z) \right]. \tag{2.57}$$

To see how this theorem works, decompose (2.57) into a pair of z transforms using (2.16), then combine the summations:

$$(z-1) X(z) = z \sum_{k=-\infty}^{\infty} x_k z^{-k} - \sum_{k=-\infty}^{\infty} x_k z^{-k} = \sum_{k=-\infty}^{\infty} (x_{k+1} - x_k) z^{-k}. \tag{2.58}$$

Because x_k is zero for all $k < 0$, we can start the summation at $k = -1$. We can also express the summation as a limit as some dummy variable goes to infinity:

$$\lim_{z \to 1} \left[(z-1) X(z) \right] = \lim_{z \to 1} \left(\lim_{p \to \infty} \left[\sum_{k=-1}^{p} (x_{k+1} - x_k) z^{-k} \right] \right) = \lim_{p \to \infty} \left[\sum_{k=-1}^{p} (x_{k+1} - x_k) \right]. \tag{2.59}$$

We can rearrange the right-hand expression to

$$\lim_{z \to 1} \left[(z-1) X(z) \right] = \lim_{p \to \infty} \left[\sum_{k=0}^{p-1} x_k - \sum_{k=0}^{p} x_k \right] = \lim_{p \to \infty} x_p \tag{2.60}$$

QED.

Use of the final value theorem is not limited to the actual final value. If you have a signal that eventually becomes a steady ramp (or parabola, or some higher exponent of time) you can find what value of velocity or acceleration it settles to by using the difference operator found in (2.51) and (2.52). By applying this

---

[7] This theorem can be easily extended to any x_k that is zero as k goes to negative infinity.

difference operator an appropriate number of times, you can change a position signal to velocity or acceleration and find the steady-state value using the final value theorem.

Initial Value Theorem

The initial value theorem states that for any signal x_k that is zero for all $k < 0$ and has a z transform $X(z)$, the value of x_k at k = 0 is equal to

$$x_0 = \lim_{z \to \infty} X(z). \tag{2.61}$$

Referring to (2.16), for a signal with $x_k = 0$ for all $k < 0$ we have

$$X(z) = \sum_{k=0}^{\infty} x_k z^{-k}. \tag{2.62}$$

If we take the limit of (2.62) as z goes to infinity, we get

$$\lim_{z \to \infty} X(z) = \lim_{z \to \infty} \left[\sum_{k=0}^{\infty} x_k z^{-k} \right]. \tag{2.63}$$

But as z goes to infinity, all of the z^{-k} terms drop out except for $k = 0$, so we get

$$\lim_{z \to \infty} X(z) = \lim_{z \to \infty} \left[\sum_{k=0}^{\infty} x_k z^{-k} \right] = x_0. \tag{2.64}$$

QED.

Notice that it is extremely rare to find a z transform of a signal that is not in the form of a ratio of polynomials (so rare, in fact, that you won't see it in this book!). Consequently, the initial value theorem can be applied by just looking at the leading term in the numerator and denominator—if the numerator polynomial is of the same order as the denominator and the denominator's leading coefficient is 1, then the initial value is simply the coefficient of the leading term of the numerator. If the numerator polynomial is of lower order than the denominator then the initial value is 0, and if the numerator polynomial is of greater order then the initial value theorem does not apply.

Summary of Z Transform Properties

Table 2.2 summarizes the important properties of the z transform.

Property	Comments	
$Z[kx_n] = kX(z)$	k must be a constant.	(2.65)
$Z[x_{1n} + x_{2n}] = X_1(z) + X_2(z)$	(2.65) and (2.66) together imply that the z transform is a linear operation.	(2.66)
$Z[x_n - x_{n-1}] = \frac{z-1}{z}X(z)$	This is called a "first difference"—it's the discrete-time equivalent to a differential.	(2.67)
$Z\left[\sum_{m=0}^{n} x_m\right] = \frac{z}{z-1}X(z)$	A summation is the discrete-time equivalent to an integral.	(2.68)
$Z[x_{n-1}] = \frac{X(z)}{z}$	This is the unit delay—it's the fundamental operation in discrete-time analysis.	(2.69)
$\lim_{n\to\infty} x_n = \lim_{z\to 1}\left[\left(\frac{z-1}{1}\right)Z[x_n]\right]$	This is called the "Final Value Theorem" and is only valid if x_k settles to a finite value.	(2.70)
$\lim_{k\to 0} x_k = \lim_{k\to\infty} Z[x_k]$	This is called the "initial value theorem" and can be verified by simple polynomial division.	(2.71)

Table 2.2 z Transform properties

2.6 Transfer Functions

Up to this point we have gone into great detail on the use of the z transform for analyzing signals, but we have not applied it to systems. If the z transform were only useful for analyzing signals it would probably not be worth the time and effort required to understand it. It is not limited to just signals, however. The z transform turns out to be a fine tool for analyzing general system behavior, specifically for being able to predict how a system will respond to inputs in general.

Finding the Transfer Function

In section 2.2 we solved a linear shift invariant difference equation "the hard way." The z transform can be used not only to solve specific difference equations, but to find general solutions to any difference equation that pertain to a particular linear shift invariant system.

Take a system as described by the generic difference equation in (2.9), which I'll repeat here:

$$x_k = \sum_{n=0}^{N} b_{N-n} u_{k-n} - \sum_{n=1}^{N} a_{N-n} x_{k-n} . \tag{2.72}$$

As long as the system remains linear and shift invariant the system behavior is always described by (2.72), no matter what its input signal may be.[8] To find a description of this characteristic system behavior one only needs to take the z transform of (2.72), taking the time shifts into account. This is permissible because the z transform is linear. By applying (2.46) we get:

$$X(z) = \sum_{n=0}^{N} b_{N-n} U(z) z^{-n} - \sum_{n=1}^{N} a_{N-n} X(z) z^{-n} . \tag{2.73}$$

This can be rearranged and multiplied on both sides by z^N to give

$$\frac{X(z)}{U(z)} = \frac{b_N z^N + b_{N-1} z^{N-1} + \cdots + b_0}{z^N + a_{N-1} z^{N-1} + \cdots + a_0} . \tag{2.74}$$

Now any time that we have a specific input signal *u* going into a system and we want to know what the output signal *x* is going to be, we can simply take the z transform of *u* and multiply it by the right-hand side of (2.74). The result will be the z transform of *x*. This means that the expression captures the entire character of the system: from the perspective of the z transform nothing more needs to be said about it.

---

8 We are carefully ignoring the fact that a system that may normally behave in a linear manner can be driven into nonlinear operating regions by certain (usually large) inputs. One needs to take care with this issue, but it must be done outside the context of the z transform, so we'll treat it in a later chapter.

We can derive an input-output relationship for any system that is amenable to analysis using the z transform. Because this relationship describes how the input signal to the system is transformed into the output signal, it is called the system's transfer function. It is usually written as a single function of *z*, like so:

$$H(z) = \frac{b_N z^N + b_{N-1} z^{N-1} + \cdots + b_0}{z^N + a_{N-1} z^{N-1} + \cdots + a_0}. \tag{2.75}$$

System transfer functions are denoted just as if they were signals, with a capital letter and a *z* in parentheses. This can be a bit confusing, so it's important to remember that the transfer function of a system is an operator that changes an input signal into an output signal. This confusion about the nature of a transfer function is compounded by the fact that if we apply a unit impulse to the system the z transform of the resulting signal will be exactly equal to the system transfer function. Because of this you will often see $H(z)$ referred to as "the system impulse response," though it would be more proper to only refer to h_k by that name.

Whenever you have a system transfer function and a z domain input signal you can find the z domain response by simple multiplication. For instance, with a transfer function $H(z)$ and an input signal $U(z)$ you can find the output signal $X(z)$ by

$$X(z) = H(z)U(z). \tag{2.76}$$

This means that by using the z transform you reduce the process of solving the difference equation that describes the system to a simple multiplication of ratios of polynomials. This follows directly from the notion that the transfer function is the ratio between the input and output signals, and it provides the z transform with much of its utility.

Example 2.4 Using Transfer Functions

Figure 2.1 shows a motor and geartrain that we might use as a plant in a servo system. It is driven with a voltage, and the position of the output gear is read with a potentiometer. Ignoring any nonlinearities in the system, its behavior is described by (2.77).

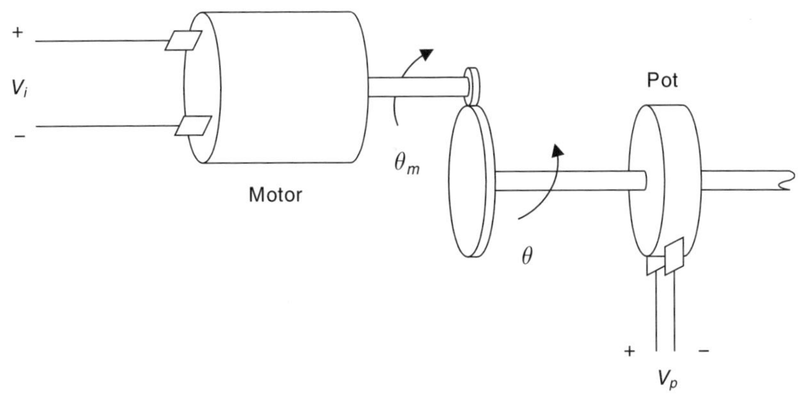

Figure 2.1 A voltage driven motor and gear train

$$x_k = (a+1)x_{k-1} - ax_{k-2} + bu_{n-1} \tag{2.77}$$

1. Find the transfer function of the motor/gear combination.
2. Use the transfer function to find the motor/gear step response assuming $a = 0.8$ and that $b = 0.2$.

   To find the transfer function, apply (2.46) as appropriate to get

$$X(z) = X(z)(1+a)z^{-1} - X(z)az^{-2} + U(z)bz^{-1}, \tag{2.78}$$

and simplify to

$$X(z)\left[1-(1+a)z^{-1}az^{-2}\right] = U(z)bz^{-1}, \tag{2.79}$$

$$H(z) = \frac{X(z)}{U(z)} = \frac{bz}{z^2-(1+a)z+a} = \frac{bz}{(z-1)(z-a)}. \tag{2.80}$$

With (2.80), if we know the z transform of u we can find the z transform of x very easily by plugging it into the right-hand side of the equation. Then we could find the inverse z transform of the result to find out what x is in the time domain.

The step response of a system is its reaction to a step function (2.38) on its input. Multiplying $H(z)$ by the unit step gives us

$$X_{step}(z) = H(z)\frac{z}{z-1} = \frac{bz^2}{(z-1)^2(z-a)}. \tag{2.81}$$

Plugging in the values $a = 0.8$ and $b = 0.2$ and performing a partial fraction expansion we get

$$X(z) = \frac{0.2z^2}{(z-1)^2(z-0.8)} = \frac{z}{(z-1)^2} - \frac{4z}{z-1} + \frac{4z}{z-0.8}. \tag{2.82}$$

By using the transform pairs in Table 2.1 we get x_k:

$$x_k = (k - 1 + 0.8^k)u(k). \tag{2.83}$$

As a final check we examine (2.83) and verify that it is the equation for a signal that starts at zero then ramps up its velocity until it has reached a steady velocity where it runs until the end of time. This matches our expectations for our motor/gear system.

2.7 Stability in the Z Domain

Normally when you are designing a closed-loop control system you want the output to follow the command, and you *don't* want the system to be volunteering its own behaviors. When a system has the characteristic that for an unchanging input its output also slows down and stops at some steady value, and for any finite, bounded input its output is also finite and bounded it is said to be "stable." Designing systems that will remain stable while achieving suitable performance goals is one of the central pursuits in control systems engineering.

When a system is linear and shift invariant its stability properties can be calculated fairly directly from information in the z domain.

Characteristic Polynomial

While (2.76) looks very simple, it points up a very important and useful property of linear systems: all of the elements in the response of the system to an input come from the system transfer function or from the input signal. You can see this in the partial fraction expansion of the output signal: each component of the output signal either results from a component of the input signal or from a characteristic of the system itself; there are no other components to the output signal.

This is a very important property, so I'll restate it: the roots of a transfer function's denominator polynomial will nearly always make their presence known in the output of the system; the only time they won't is if the signal has a numerator polynomial with a matching root. Moreover, the output signal will contain *no* components that are *not* in the input signal or transfer function. This means that if you are interested in the general behavior of the system to the response of an arbitrary signal you don't have to try out all possible signals on your system,[9] you only need to factor the denominator and look at the value of its roots.

The usefulness of the transfer function denominator in predicting system behavior is so important, in fact, that control engineers refer to it as the system's "characteristic polynomial." All of the information about a system's stability, and much about how rapidly it will respond to inputs, is contained in the roots of its characteristic polynomial, which are called the *poles* of the system.[10]

Each pole of the transfer function will generate a component of the system response that shrinks (or grows) as the absolute value of the pole. This means that if you have a pole at $z = d$ there will be a component of the response that goes as $|d|^k$. This applies to any pole whether it is a single real pole or one of a complex pair. The locus of points where the absolute value of z is unity ($|z|=1$) is thus the boundary between values of z for which the system has stable poles and values of z for which the system has unstable poles; it is called the "unit circle."

As a consequence of the response following the poles of the system, one can make some rough predictions of the system behavior simply by knowing the locations of the poles. A system with poles close to $z = 0$ will settle rapidly; a system

---

9 This is not the case with nonlinear systems; we'll investigate that idea later in the book.

10 Because the magnitude of the transfer function as z varies over the complex plane goes to infinity at each denominator root, as if it were a circus tent with a pole at that value.

with poles that are close to the unit circle will settle slowly; a system with poles right on the unit circle may have a sustained response, or with certain bounded inputs it may have an ever-growing response; finally, systems with poles outside the unit circle will have a response that grows exponentially.

Because a system's response is determined by the locations of its poles and zeros, it is common to plot the system poles and zeros with the unit circle for reference. Poles are plotted as 'X', zeros as 'O', and multiple poles or zeros are indicated by a number.

Example 2.5 A Pole-Zero Plot

A system has the transfer function

$$H(z) = \frac{z^3}{(z-0.8)(z-0.9-j0.17)(z-0.9+j0.17)}. \tag{2.84}$$

Construct its pole-zero plot, find the components of its response, and make an estimate of how long it would take for its response to settle out to within 1% of its greatest value after a change in its input.

The numerator of the transfer function has a triple zero at $z = 0$, and the pole locations can be seen by inspection. The result of plotting these is shown in Figure 2.2.

The system response has components at each of the three poles:

$$y(k) = y_T(k) + A(0.8)^k + B(0.9+j0.17)^k + B^*(0.9-j0.17)^k \tag{2.85}$$

where y_F is the forced response with components from the input signal. Note that B is a complex number and that B^* is its complex conjugate. The complex poles can be reduced using Euler's identity and the signal shown as

$$y(k) = y_T(k) + A(0.8)^k + C(0.916)^k \cos(\phi + 0.17k). \tag{2.86}$$

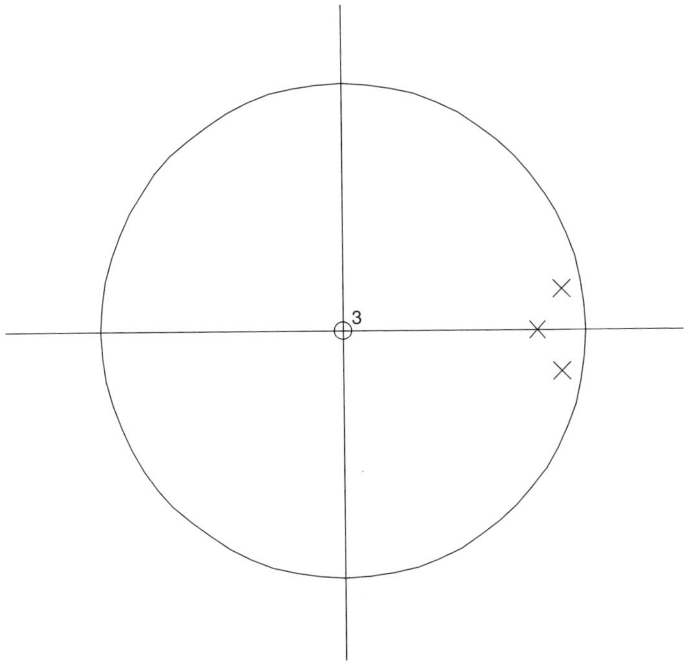

Figure 2.2 A pole-zero plot

The signal component that settles the slowest is the cosine wave; it will be 1% of its greatest response when

$$(0.916)^k = 0.01, \tag{2.87}$$

this will happen when k is greater than 52.

2.8 Frequency Response

If you excite a system with a sine wave and look at the amplitude and phase shift of the resulting output, you can characterize the system without ever having to extract a "writable" transfer function.

Say that you have a system with the transfer function $H(z)$. If you were to excite this system with the signal $x(k) = \mathrm{u}(\mathrm{k}) \sin(\theta k)$ then the response of the system would consist of two parts: a homogeneous response $y_h(z) = A \sin(\theta k + \phi)$ and a nonhomogeneous response whose character will depend on the characteristics of $H(z)$. Calculating this response given $H(z)$ can be done using partial fractions culminating in an exercise to extract the amplitude and phase, but it can be simplified by noting that when you use Euler's identity the excitation can be broken down to

$$x(k) = \sin(\theta k) = \frac{1}{2}\left(e^{j\theta k} - e^{-j\theta k}\right). \tag{2.88}$$

Similarly, the response can be extracted

$$y(k) = A\sin(\theta k + \phi) = \frac{A}{2}\left(e^{j(\theta k+\phi)} - e^{-j(\theta k+\phi)}\right). \tag{2.89}$$

A bit of fiddling with partial fraction expansions will show that

$$H\left(e^{j\theta}\right) = Ae^{j\phi}. \tag{2.90}$$

In control systems and signal processing A is called the system *gain*, and ϕ is called the system *phase shift* (or just phase). Gain is usually expressed in *decibels*, which is commonly abbreviated dB. The term comes from electronics and is nominally a power gain; 10dB of gain is a 10 times increase in power level from the input to the output. When used with signals it is assumed that the signal is a voltage or current driving a resistance, so the power level is proportional to the signal level squared. In that case the level in dB of signal y is $20 \log_{10}(y)$.

Because signal levels are measured in dB, a gain can be expressed in dB as well. If $y = A\,x$ then $20 \log_{10}(y) = 20 \log_{10}(A) + 20 \log_{10}(x)$. Thus, if we say that the system with transfer function H has 20dB of gain at some frequency, we mean that $A = 10$ at that frequency.

When you express a transfer function in terms of its gain and phase as a function of frequency, the result is called a frequency response. So far we've used a system's transfer function to find its frequency response. You can, in fact, define the whole system's behavior using only its gain and phase response. Then you can use the frequency response to design a controller, without ever referring back to the z-domain model of the system.

In fact, you don't even have to *have* a z-domain model of your plant to use these design techniques—you can measure the system frequency response at a number of different frequencies and use this measured response for your controller design without ever needing a z-domain model of the plant. All you need is a setup that lets you stabilize the plant long enough to get your measurement. This is an enormous aid in practical control system design. It is often difficult if not impossible to get an adequate system model in the z domain, yet by taking a fairly simple set of measurements you can sidestep the whole issue and design a high-performance control system without ever generating an explicit z-domain model.

If you compute or measure a system's gain and phase over a range of frequencies and plot the resulting information with a logarithmic frequency axis the result is called a *Bode plot*. We will use Bode plots extensively in Chapter 5; for now, I'll just give a couple of examples.

Take the case of a system with a sampling rate of 100Hz whose transfer function is

$$G_{ol}(z) = 0.2\frac{(z-0.97)(z-0.98)}{(z-1)^2(z-0.995)}. \tag{2.91}$$

Figure 2.3 shows a Bode plot of the transfer function in (2.91).

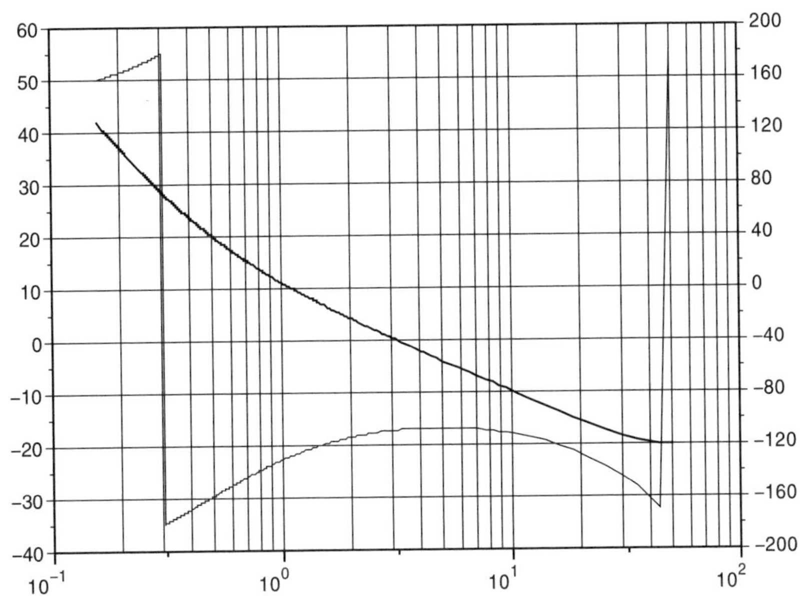

Figure 2.3 Bode plot of (2.91)

When you have frequency response information at hand, any computations that you do with transfer functions can still be performed at any specific frequency—simply replace a transfer function with its value at that frequency, and the problem is reduced to simple complex arithmetic.

Take the case of the system described by (2.91). If this system were the controller and plant in a closed-loop system, the system's closed-loop gain would be

$$G_{cl}(z) = \frac{G_{ol}(z)}{1 + G_{ol}(z)}. \tag{2.92}$$

If we had the transfer function spelled out for us as we do in (2.91) we could compute the transfer function of the closed-loop system in the z domain. If all we had was the Bode plot in Figure 2.3, however, we could not get this transfer function. We *could* still get a Bode plot of the system by computing (2.91) at each frequency point, in which case we'd get the Bode plot in Figure 2.4.

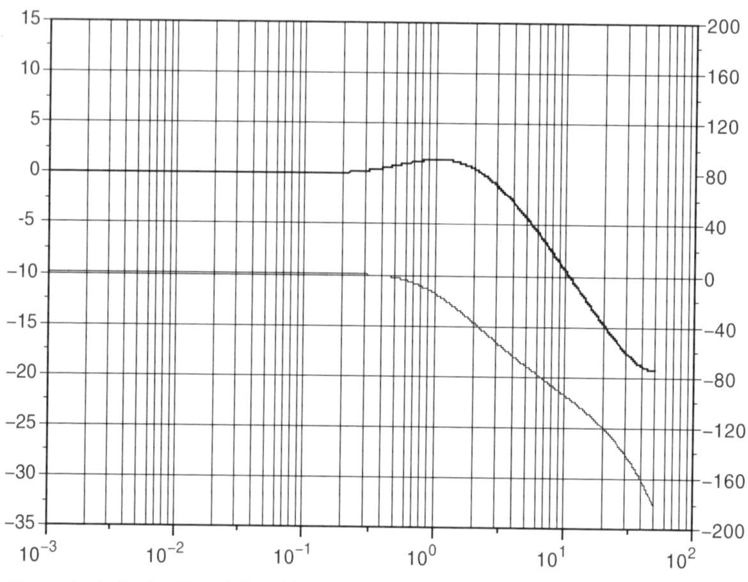

Figure 2.4 Bode plot of closed loop system

What all this frequency-response discussion means is that if you have a measured plant response in hand you can perform any analysis or design tasks necessary using transfer functions for your (known) controller and your measured data for your (unknown) plant. Such a hybrid approach is often invaluable in a system with a digital controller—since you have excellent control over the controller transfer function, you only need the measured plant response to complete the picture. If your plant is well-behaved enough, you may even be able to perform all your design from one initial measurement of the plant response.

2.9 Conclusion

This chapter has presented the essential information about the z transform that will be used for the balance of this book. It has defined the z transform, has shown how the z transform can be used to solve difference equations, how it can be used to represent systems, and how you can extract information from system transfer functions without getting lost in the details.

Finally, we have touched on frequency response analysis and how it can be used to bring measured system response data into the realm of z-domain analysis, without actually needing to have a z-domain model of the system.

3

Performance

It is not enough to be able to describe how a system works. It is also necessary to design systems that will do what they need to do. In order to design a system to a given performance goal, one must first be able to describe the system requirements in a consistent and well-known manner.

There are performance measures for control systems that are generally used in industry because they are easy to state in understandable terms, yet they are easy to relate to a system's description in the z domain. This chapter introduces these commonly used measures of control system performance in the time and frequency domains, and relates them to one another and to the system's z-transform description.

3.1 Tracking

The goal of nearly all control systems is, in some way, to follow a commanded trajectory. This trajectory may be varied from outside the system such as one would see in a servo control system, it may be a constant steady value such as one would see in a process control system, or there may be a desire for the system to follow some predetermined trajectory autonomously, in which case the system design can often be divided into an "executive" controller generating the target trajectory, followed by a servo controller.

An obvious and powerful measurement of a control system's performance is how well it follows its commanded input in the time domain. How soon it starts moving, how quickly it comes to its target, and how long it takes to stop moving once it comes close, are all important measures of performance.

Step Response

With the exception of a constant setpoint, the most basic command that a system can be given is a step command: the system has been operating at one setpoint, and suddenly we want it to operate at another. Ideally, the system would follow this change in command exactly, but of course in reality it does not.

Figure 3.1 shows how the various performance parameters of the step response are commonly defined. There are five elements to the definition: the settling time (t_s), the rise time (t_r), the delay time (t_d), the overshoot, (a_v) and the steady-state error (a_{ss}). As usually defined, the settling time is the amount if time it takes for the response to get within a certain percentage (often 10%, but this can vary) of the commanded (or final) value. The overshoot (if any) is the percentage by which the response exceeds the commanded or final value, the rise time is the amount of time it takes to get to some percentage of the final value (usually 90%), the delay time is the amount of time it takes for the response to reach some percentage of final (usually 10%), and the steady-state error is the difference between the commanded value and the actual response.

These parameters are commonly used, but their definitions can vary. It is best when specifying system performance or when designing a system to specifications, that these definitions be clarified. One should say (or look for) fully qualified statements like "settling time (to within 5% of the final value)" or "overshoot

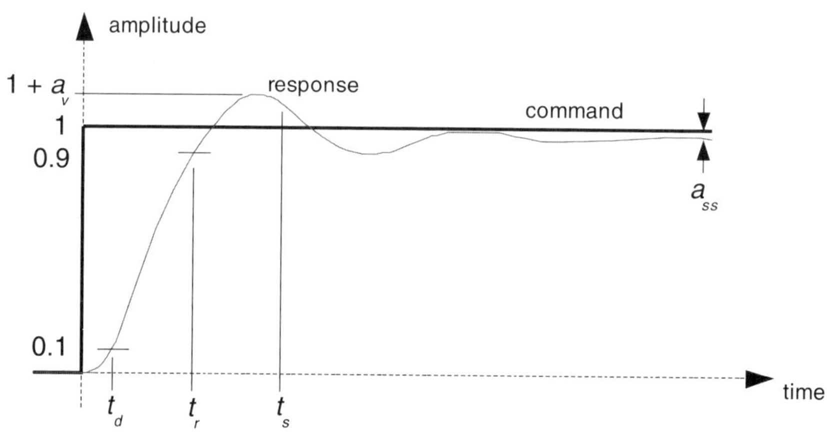

Figure 3.1 Elements of the step response

(above commanded value)," etc. Ambiguity when specifying should be avoided, and ambiguity of specification should be questioned, or at least noted in one's design documentation.

If the system can be described as a simple first- or second-order system, then the step response parameters can be derived mathematically. Even though few real systems are so simple, these responses can be used to get a quick idea of how an actual system will respond. In the discussion to follow, I am going to assume that a_{ss} is zero, because this is easy to achieve,[1] and it is often assumed to be true. The following results can all be extended with a nonzero a_{ss} if desired.

First-Order System

Take the case of a first-order low-pass system with the transfer function

$$T(z) = \frac{1-d}{z-d}. \tag{3.1}$$

This system will have the step response

$$R_{step}(z) = \frac{(1-d)z}{(z-1)(z-d)} \Leftrightarrow r_{step}(k) = 1 - d^k. \tag{3.2}$$

With a first-order system the delay time is a trivial parameter, but it can be calculated:

$$1 - d^{t_d} = 0.1. \tag{3.3}$$

This equation gives the time as a real number, but we are dealing with controllers that operate in sampled time. For a plant that has a sufficiently low-pass nature (3.3) is an accurate approximation. A more conservative estimate would be to round up to the next higher integer:

$$k_d = \text{ceil}\left(\frac{t_d}{T_s}\right) = \text{ceil}\left(\frac{\ln(0.9)}{\ln(d)}\right). \tag{3.4}$$

---

1 See the development of integrating compensators in Chapter 6, section 6.3.

If the system is being sampled fairly slowly in comparison to its response, and if the exact number is important, it will be necessary to take the continuous-time nature of the plant into account to find the exact delay time (or any of the other times derived here).

The derivation of the rise time is very similar to that for the delay time; only the value at the instant that the rise time is met needs to be changed:

$$1 - d^{t_r} = 0.9 \cdot \tag{3.5}$$

$$k_r = \text{ceil}(t_r) = \text{ceil}\left(\frac{\ln(0.1)}{\ln(d)}\right) \tag{3.6}$$

Because a first-order low-pass filter has a monotonically increasing response that approaches the final value asymptotically, the settling time will be the same as the rise time, and there will be no overshoot.

It is often useful to state the "time constant" of a first-order system. The term "time constant" comes from continuous-time systems, and it is defined by expressing the system response in terms of the natural exponent:

$$R_{step} = 1 - d^k = 1 - e^{k\ln d} \, . \tag{3.7}$$

The time constant is the time required for the exponent to reach –1:

$$\tau = -\frac{T_s}{\ln d}, \tag{3.8}$$

where T_s is the sampling time, and τ is the time constant. For all but very low sampling rates, the time constant is useful, because a number of system characteristics scale directly with the time constant:

- The amount of time for the system to get to within 37% of the final value $(1-e^{-1})$ is one time constant.
- In five time constants, the system is within 1% of its final value $(1-e^{-5} = 0.007)$.
- Delay, rise and settling times scale with the time constant

Second-Order System

With the step response of a second-order system, the delay and overshoot become important behaviors, and we now have to pay attention to the three parameters associated with them.

Second-order systems can be roughly characterized by their poles, which can both be located on the real axis, or which can occur in a complex conjugate pair. Complex conjugate pairs are called "resonant poles" or "resonant pairs," and are characterized by their damping ratio. When a pole pair occurs in complex conjugate you can express them as

$$z = e^{-a-j\theta}, \tag{3.9}$$

which leads to a system response like

$$y_k = Ae^{-a+j\theta} + A^*e^{-a-j\theta} + \cdots \tag{3.10}$$

(remember that the notation A^* indicates the complex conjugate of A). If the Euler identity is used then (3.10) becomes

$$y_k = 2Ad^k\cos(\theta k + \phi), \tag{3.11}$$

where d is taken to be e^{-a} and ϕ is the polar angle of A.[2]

The behavior of the pole pair depends on the values of a and θ: The parameter a determines how quickly the envelope of the oscillation dies down, and θ determines how quickly the pole pair oscillates. While these measures are very useful by themselves, it is also useful to characterize the pole pair in terms of its damping ratio and natural frequency.[3]

The natural frequency of a pole pair is a rough measure of the overall speed of the pole pair response, and is defined as the absolute value of the exponent in (3.9):

$$w_n = \sqrt{a^2 + \theta^2}. \tag{3.12}$$

---

[2] For example, $\phi = \tan\left(\frac{Im(A)}{Re(A)}\right)$.

[3] The concept of a natural frequency is much more clear and direct when you are working on a continuous-time system in the Laplace domain, but it is still useful here, even if one must struggle to bring the concept into the z-domain.

The damping ratio of a pole pair compares the "useful" speed of the pair's response (in other words, how fast it decays) with the oscillatory speed. It has the most meaning when it is allowed to range from 0 to 1. It is written as ξ (the Greek letter xi) and it is defined as

$$\xi = \frac{a}{w_n} = \frac{a}{\sqrt{\theta^2 + a^2}}. \tag{3.13}$$

Inspecting (3.9) and (3.13) you can easily see that a pole pair with a damping ratio of 1 is a double pole on the real axis, with no oscillation at all, while a pole pair with a damping ratio of 0 sits right on the unit circle, and would oscillate at the same amplitude forever, without ever settling down.

From (3.12) and (3.13) one can express a and θ in terms of the natural frequency and the damping ratio:

$$a = \xi w_n \tag{3.14}$$

$$\theta = w_n \sqrt{1 - \xi^2}. \tag{3.15}$$

Contrary to appearances, a system can have a damping ratio greater than one. This comes about because when (3.15) is evaluated for $\xi > 1$ the resulting value of θ is purely imaginary; then (3.9) becomes

$$z = e^{w_n\left(\xi \pm \sqrt{\xi^2 - 1}\right)} \tag{3.16}$$

which of course describes a pair of real-valued roots.

Second-order systems can be roughly classified by how their damping ratios compare to one: a system with $\xi = 1$ is "critically damped," a system with $\xi < 1$ is "under damped," and a system with $\xi > 1$ is "over damped." We won't say much more about over damped systems, as their behavior rapidly approaches that of a single-order system with increasing damping ratio. The effect that the damping ratio has on a system as it ranges between 0 and 1, however, is of interest to control engineers.

For an example, take a low-pass system whose transfer function is

$$H(z) = \frac{\left(1 - 2d\cos\theta + d^2\right)z}{z^2 - 2d\cos\theta z + d^2}. \tag{3.17}$$

If d is taken to be e^{-a} then this system's pole locations are described by (3.9). The step response to this system can be found to be

$$y_k = 1 - d^k\cos(\theta k) + d^k\frac{d - \cos\theta}{\sin\theta}\sin(\theta k). \tag{3.18}$$

This can be restated in terms of the natural frequency and damping ratio:

$$y_k = 1 - e^{-\xi w_n k}\left(\cos\left(w_n\sqrt{1-\xi^2}k\right) + \frac{e^{-\xi w_n} - \cos\left(w_n\sqrt{1-\xi}\right)}{\sin\left(w_n\sqrt{1-\xi}\right)}\sin\left(w_n\sqrt{1-\xi^2}k\right)\right). \tag{3.19}$$

Figure 3.2 shows the y_k that results when we choose a natural frequency of 0.1 and a variety of damping ratios. Evident in this figure is the variety of different responses that arise just from a change in damping ratio. The response with a

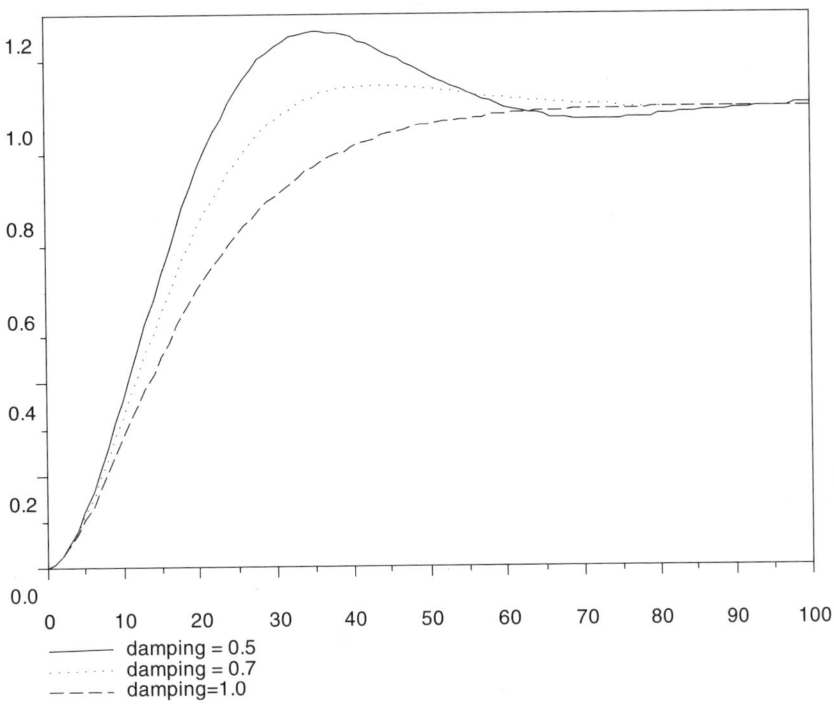

Figure 3.2 Effect of damping ratio on step response

low damping ratio starts off quickly, but doesn't slow down soon enough, and overshoots the target. The response with the high damping ratio doesn't overshoot at all (in fact $\xi = 1$ implies a double real-valued pole, and any $\xi > 1$ implies a pair of poles on the real axis). Finally, the response with a medium damping ratio has a moderate amount of overshoot and responds much faster than the high damping ratio case.

Notice the response for $\xi = 0.5$. This oscillation around the steady-state value is called *ringing*, and it gains amplitude and takes longer to settle as the damping ratio goes towards zero. Not only will a system with a low damping ratio take a long time to settle and experience a high degree of overshoot, but a great deal of ringing in a closed loop system is usually an indication that the control loop is on the verge of instability.

Figure 3.3 shows how the damping ratio affects the rise and settling time of the system. The rise time decreases with decreasing ξ, but because of the ringing the settling time is worse for low damping ratios than it is for higher ones.

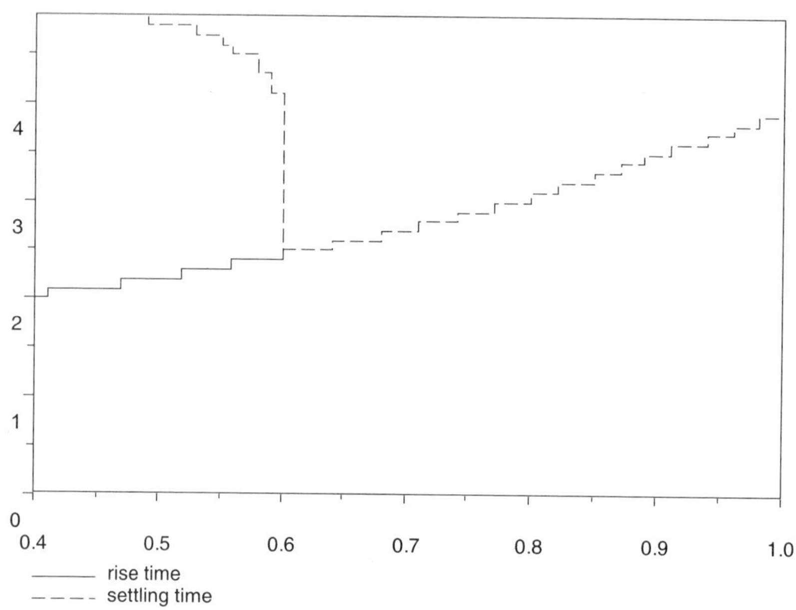

Figure 3.3 Rise and settling times as a function of ξ

Figure 3.4 shows the amount of overshoot as a function of the damping ratio. Overshoot becomes more severe as the damping ratio is decreased, reaching 100% with a damping ratio of 0.

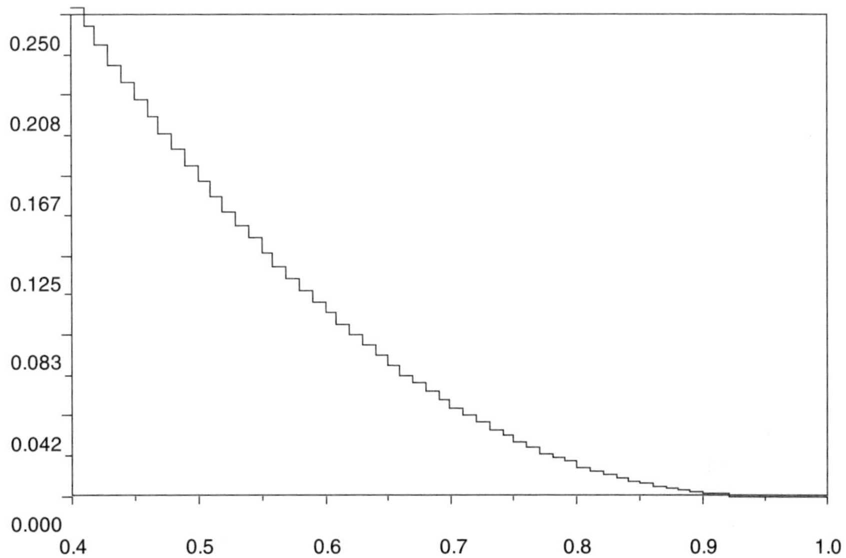

Figure 3.4 Overshoot as a function of ξ

In a continuous-time system, the shape of the response is solely dependent on the damping ratio and natural frequency. This is also the case with the sampled time system, but with a wrinkle: in a sampled time system, one can vary the sampling rate. One would like to think that one could scale the natural frequency with the sampling rate and get a consistent response; Figure 3.5 shows what happens if we attempt this: the rise and delay times are normalized for natural frequency for a system with $\xi = 0.7$ as the natural frequency varies from 0 to 1. While the rise and delay times do stay within the same order of magnitude they certainly don't stay constant!

The situation is slightly better when overshoot is considered; Figure 3.6 shows the overshoot of the filter as a function of the sampling rate. While it does vary visibly as the sampling rate varies, it remains constant enough to make general statements.

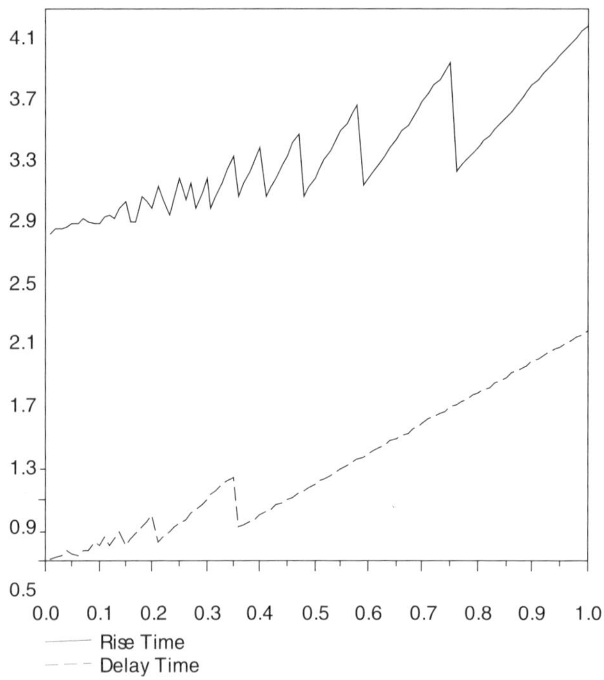

Figure 3.5 Rise and delay time as a function of sampling rate

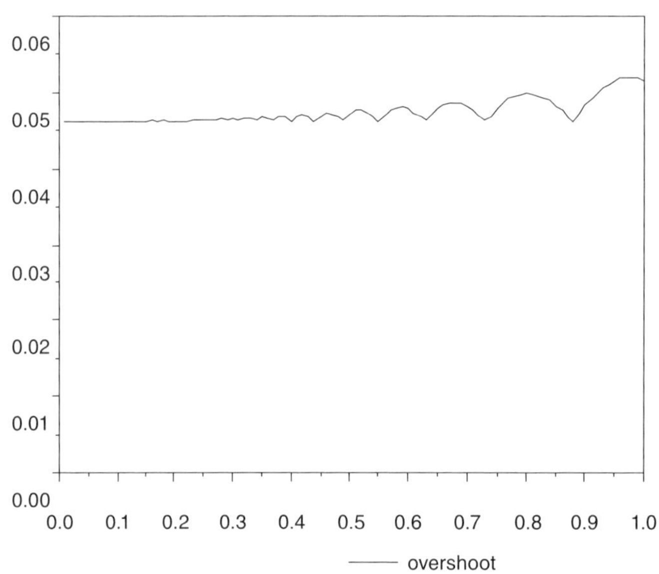

Figure 3.6 Overshoot as a function of sample time

Higher-Order Systems

Even given the variation in the system response with varying sample time, one can still draw quite a number of useful conclusions from the second-order case, and one can get a very good idea of how the system is going to respond just from the damping ratio.

In general, when a system has three or more poles that contribute to the response, it becomes impractical to try to make any generalizations at all. Sometimes you can see from performing partial fraction expansion, or from the structure of the system, that the system has a first-order pole (or a resonant pole pair) whose response dominates the step response of the system. In this case we call this pole (or pole pair) a *dominant* pole (or pole pair). When this happens, we can treat the system as a first- or second-order system for the purposes of preliminary work.

This technique is very handy, but it is important not to always assume that there is a dominant pole or pole pair. Take the case of the third-order system whose response is shown in Figure 3.7. This system might appear to have a first-order dominant pole, but if you look at its response, you can see that it has a resonant pole pair that is quite underdamped—the "stair-step" effect in the step response is from the ringing of this pole pair. While it doesn't appear to be a bad

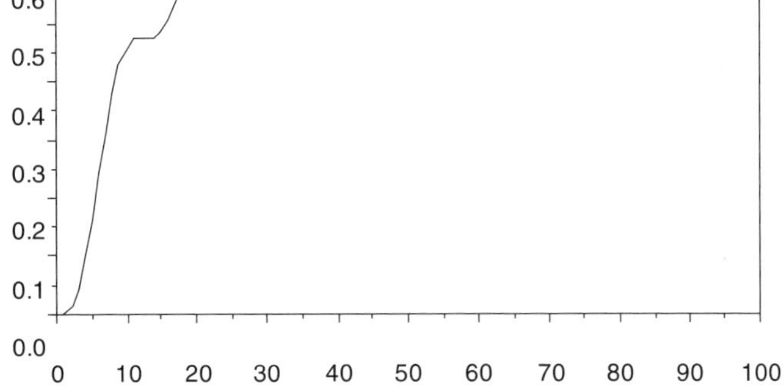

Figure 3.7 Third-order response with underdamped pole

thing in this case, such a highly resonant pole pair could cause problems if the system were to be excited by just the wrong frequency of input, and it indicates that the system may easily go unstable.

Ramp and Higher-Order Responses

System step responses are informative and useful in large part because a step input, or a step-like input, is nearly ubiquitous. When this is the case, a system's step response tells us much about its real-world behavior. If a system is designed to track a moving input, then a step response will not tell the whole story, and it is sometimes useful to characterize the system's response to a ramp, parabola or other higher-order response.

Figure 3.8 shows the response of a system to a ramp input. The figure shows that you can translate most of the parameters used for the step response onto the ramp response. Note that the rise time is fairly meaningless, but that undershoot (a_u), which was essentially meaningless for a step response, now has an important role to play.

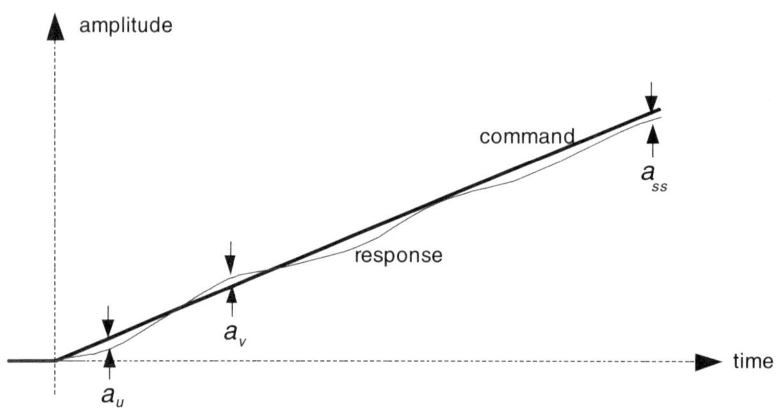

Figure 3.8 Elements of the ramp response

One very important difference between a system step response and the higher order responses is that a step response is bounded, while a higher order response is not. This means that with a stable system, the final error (a_{ss}) may not be finite; it also means that a system will never see a higher order response that goes on for an indeterminate amount of time—at some point a limit will be reached.

3.2 Frequency Response

Getting the standard time domain responses of a system is useful, but it is not always clear from a given time domain response how the system will react to continuous, perhaps random input. These sorts of inputs are often much easier to express in frequency domain terms, so it is often useful to express certain system performance criteria in the frequency domain.

Systems that are intended solely to operate on signals and which have certain frequency characteristics are referred to as "filters," because they filter out information or energy at certain frequencies while allowing others to pass. When referring to a system's frequency response characteristics, the choice of the term "system" or "filter" is one of intent. The term "filter" would certainly be appropriate for a few components on a board, and perhaps for some other system with the same transfer function. Using "filter" to describe a mechanism with the same transfer function but which is ten feet tall and has 500-horsepower motors scattered through it might be considered a misnomer, however.

Table 3.1 lists some common filter types with comments on how they are often described.

Filter Name	Plot	Comment
Low-pass	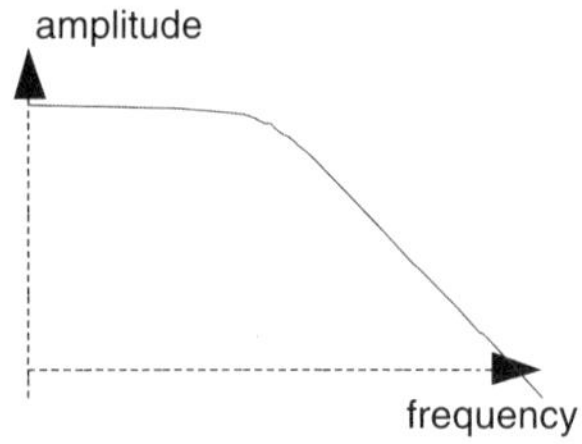	Passes low frequencies, filters out high ones. Usually described with filter order and bandwidth, often 3dB bandwidth.
High-pass	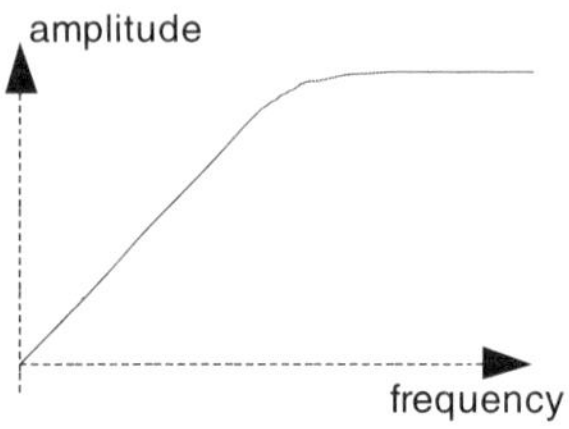	Passes high frequencies, filters out low ones. Usually described with filter order and the 3dB point, but the term "bandwidth" isn't used.
Bandpass	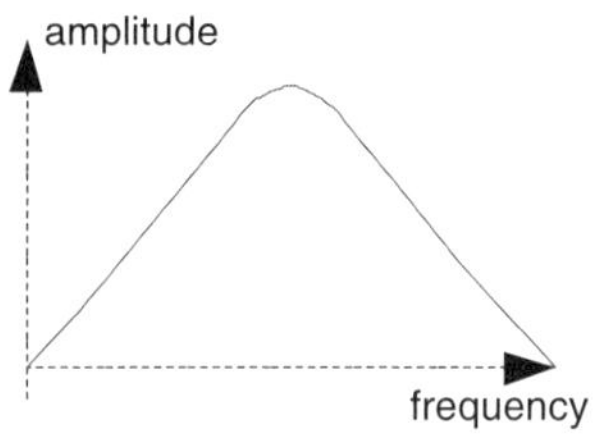	Passes a band of frequencies, not often used in control work. Usually described by center frequency, filter order and 3dB bandwidth, sometimes 6dB bandwidth. High-attenuation (40 or 60dB) bandwidth often specified, or the filter's "shape factor."
Notch	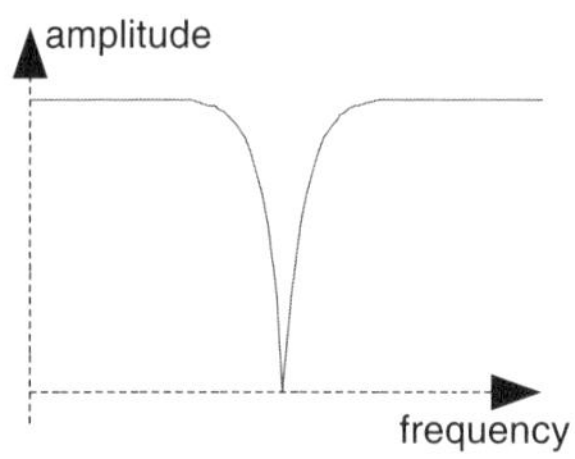	Stops a relatively narrow band of frequencies. Very useful in a very limited number of cases in control work. Described by notch frequency and 3dB bandwidth, sometimes by notch depth.

Table 3.1 Types of filters

DC Gain

The most important aspect of a closed-loop system is that it is stable. After we've assured ourselves that a system is stable, the next most important characteristic is usually its steady-state input-output relationship; that is, the relationship between the system's output to a steady input, after everything has settled. This factor is called the system's DC gain.

To find a system's DC gain, we simply use the techniques of Chapter 2 to find the system gain at zero frequency, in other words at $z = 1$. Doing so is tantamount to giving the system a unit step input, then using the final-value theorem to find its value at infinite time.

Bandwidth

There are a number of different definitions of the term "bandwidth," but all of them are a measure of the degree to which a system passes a signal from its input to its output over the frequency spectrum.

In a low-pass system, "bandwidth" means the 3dB bandwidth; that is, the frequency at which the system's amplitude response has fallen to 3dB (half the power) of its response at DC. Figure 3.9 is an amplitude plot of a simple first-order low-pass system[4] showing its 3dB point at f_o. One will occasionally see

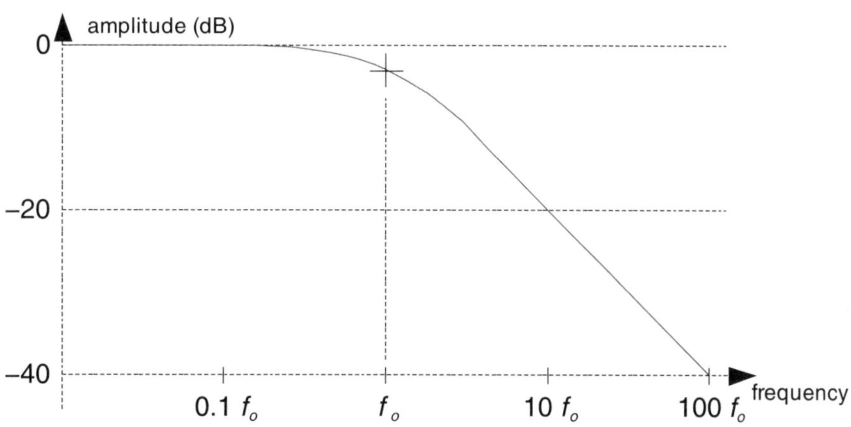

Figure 3.9 Bandwidth of a low-pass filter

4 This system's pole is very close to 1, which is why the amplitude decays in a straight line as a function of frequency. As a sampled-time system's poles recede from the unit circle, their effect on a Bode plot becomes less direct.

system bandwidths specified at the 6dB point or something even odder; if this is so, then the author of the specification should make it clear what they mean (often just saying, for example, "6dB bandwidth," is enough).

In a bandpass filter, the bandwidth is specified as the difference between the lower and upper frequencies where the response falls to 3dB (or 6dB) below the peak bandwidth. In addition a bandpass filter is specified as having a center frequency, which is usually centered between the two 3dB (or 6dB) points.

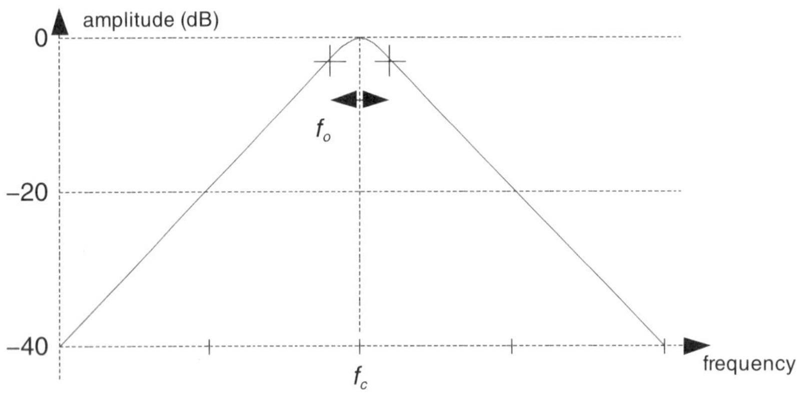

Figure 3.10 Center frequency (f_c) and 3db bandwidth (f_o) of a bandpass filter

It would be very nice if there were simple and mathematically correct generalizations that could be made about a system's time-domain response in relation to its frequency-domain response. Using the Fourier transform for discrete-time signals, one can go from a system's frequency response to its transfer function, but this procedure is not mathematically simple. What *can* be stated accurately is that if you have two systems with the same frequency response plots in the z-domain and different sampling rates then their time-domain responses will be scaled in time similar to their sampling rate scaling.

The difficulty arises because while a system with a higher bandwidth generally has a higher response speed and faster settling time, the presence of "ringing" at high or low frequencies depends on how sharply the system's amplitude response falls off, and in exceptional cases[5] on variations in a system's phase response.

5 A system with zeros outside the unit circle can have a perfectly flat amplitude response, yet have a highly resonant pole pair. Systems with such "unstable" zeros are known as non-minimum phase systems.

None the less, given a system that does not have a great multiplicity of zeros, and for which a lack of highly resonant poles can be assumed, it is generally correct to assume that the system's time-domain response will be similar in timing to a first- or second-order system with a similar bandwidth. Such an approximation will suffice for estimations of a system's performance, but this should always be checked by looking at the actual system's time response.

Phase Shift

Phase shift in a system is important for two reasons: first because the amount of tracking error in a system depends on its phase shift over frequency as well as its amplitude variations, second because (as we'll find in later chapters) the phase characteristics of a system's various pieces strongly affect the performance and stability that can be achieved by the system as a whole.

Example 3.1

A system has a sampling rate of 1000Hz and the transfer function

$$H_{lp}(z) = \frac{z^2}{(z-0.9)^3}. \tag{3.20}$$

Plot the filter's gain and phase response vs. frequency, then plot the difference between unity and the filter's magnitude response as a function of frequency, for example

$$1 - \left|H_{lp}(z)\right| \tag{3.21}$$

and the filter tracking error, for example

$$\left|1 - H_{lp}(z)\right|. \tag{3.22}$$

Figure 3.11 shows the phase and magnitude plots for the filter; notice that the phase begins showing a deviation over a decade before the magnitude.

Figure 3.12 shows the actual and apparent tracking error in the third-order low-pass filter as a function of frequency.

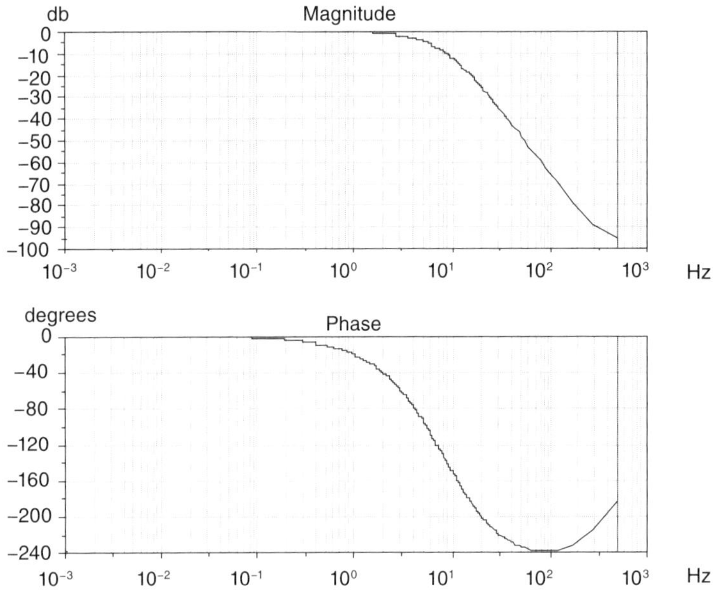

Figure 3.11 Phase and magnitude plot of third-order low-pass filter

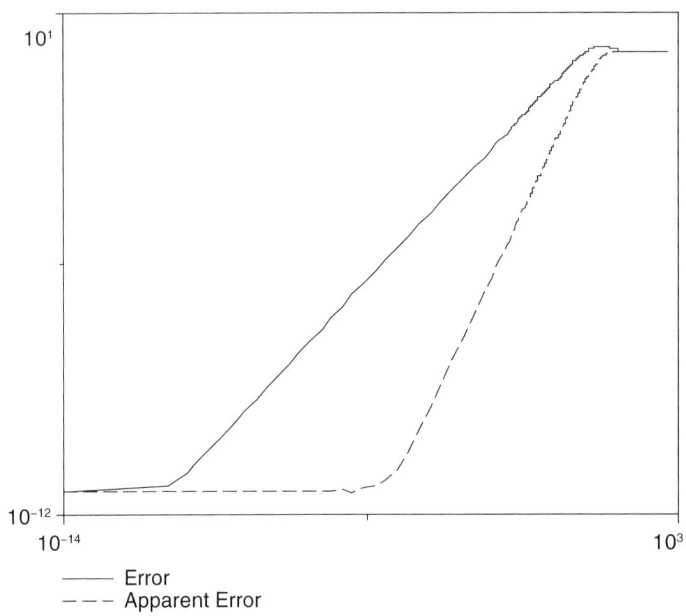

Figure 3.12 Real and apparent error of third-order low-pass filter

Notice an important aspect of this plot: the actual error, while it is the same at high frequencies, starts to rise much earlier than the apparent error that is derived from using magnitude differences alone. This is because the amount of phase change induced by the filter dominates the error at all but the very highest frequencies where the error is apparent.

3.3 Disturbance Rejection

Not all control systems are designed to track a target. There are probably as many control systems whose sole reason to exist is to hold some quantity at a specific setpoint as there are control systems designed to track a command. These control systems are called "regulators," and while their step response can be an important indication of their performance, their ability to reject disturbances has a far greater impact on their ability to operate successfully.

Systems are always affected by external, unintentional or undesired effects. Few (if any) systems can be built so that they are completely free from undesired responses arising from external influences. Control system designers refer to these external influences as "disturbances." Disturbances can take nearly any form, and are practically innumerable. Environmental changes such as temperature, sunlight or humidity can act directly or indirectly as inputs to a system; a motion-control system may be mounted onto a platform that is itself moving, a system may be subject to electrical noise, and so on.

Figure 3.13 shows a simple block diagram of a control system that suffers from a disturbance input. In this case u is the intended command input, and u_d is the disturbance input. We will use this system as our model for the following discussion.

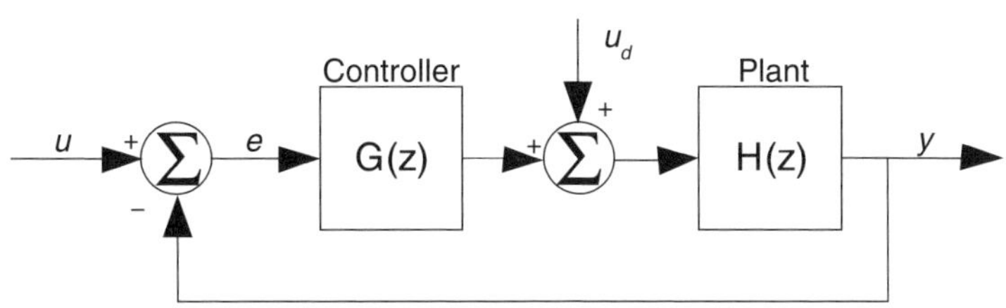

Figure 3.13 A system with a disturbance

An ideal control system would entirely eliminate its plant's response to unwanted inputs. Unfortunately, this is not possible: a control system can strongly affect a system's response, but there are fundamental limits[6] to the nature and amount of disturbance rejection that a controller can bestow on a system.

This leaves us with the necessity to define the amount of output disturbance that is acceptable given a certain input disturbance. The two ways of expressing output disturbances are, not surprisingly, derived from the two ways of expressing tracking performance: we can define our disturbance rejection in the time domain with maximum deviation, DC residue and settling times, or we can define our disturbance rejection in the frequency domain in terms of input-output gain.

Time Domain

Figure 3.14 shows how a system might respond to a step disturbance. At the leading edge of the disturbance, the system output rises to some peak value, then it settles down to a steady-state.

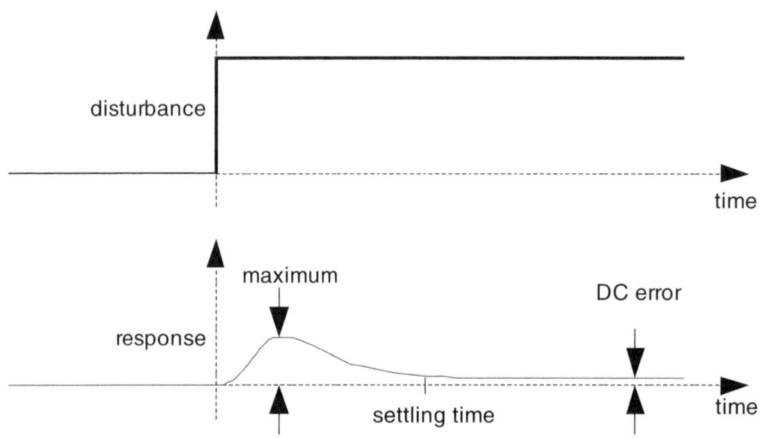

Figure 3.14 An example disturbance response

[6] Bode's Sensitivity Integral demonstrates, in fact, that there is a certain "conservation of sensitivity" and that while one can push a plant's sensitivity into frequency bands where it may have less harmful effects, one cannot reduce the total sensitivity.

Normally, the salient points of the disturbance response will be the maximum amplitude it reaches, how long it takes to settle, and any residual DC error that it leaves. Other parameters of the disturbance response that may be of interest would be its amplitude in one direction or another (in other words, the direction of the disturbance response may matter), or possibly the total area of the response, or the total energy of the error response.

Frequency Domain

As with the intended response of a system, the disturbance response of a system can be shown as a frequency response plot in magnitude and (if desired) phase. A system's disturbance response plot will often take the form of a bandpass filter—at low frequencies, the control system actively controls the system output, at high frequencies most plants act as low-pass filters, and at intermediate frequencies the plant is still passing a substantial amount of disturbance while the controller doesn't have enough authority to affect the plant behavior in any beneficial way.

3.4 Conclusion

This chapter covered the common methods for defining the performance of a control system. It has shown a few ways that the system characteristics in the z domain could be used to predict how well a system would meet a given set of specifications. This information will be useful in the early phases of a system design when you are determining what is necessary to make the system successfully meet its real world design goals.

4

Block Diagrams

Control systems engineers use block diagrams extensively in system analysis and design. Block diagrams provide two major benefits to the control system engineer: they provide a clear and concise way of describing the behavior and structure of the system, and when the block diagram "language" is limited appropriately, they provide formal methods for analyzing system behavior.

This chapter will cover methods of describing a system using block diagrams, a block diagram "language" that I have found to be very clear and detailed, some examples of other block diagramming "dialects" that you'll see in related literature, and finally some methods for analyzing system behavior using block diagrams.

4.1 The Language of Blocks

Figure 4.1 shows a block diagram. It is probably the world's simplest block diagram, but it is complete, nonetheless. The signal u feeds into a system ("Thing" in the diagram), which transforms it into the signal y. The behavior of "Thing" is unspecified, so it could denote any signal processing function that accepts one signal and transforms it into another. What *is* specified in this block diagram is a single input signal, a single output signal, and the dependence of the output signal on the input signal and its past values, and possibly on the initial conditions of the block "Thing" at some specified point in time.

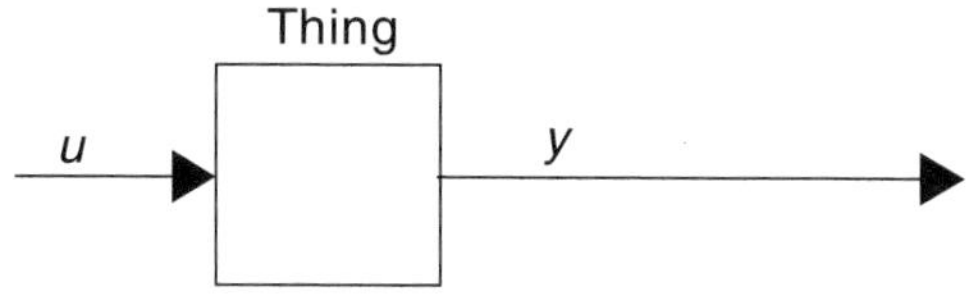

Figure 4.1 A simple block diagram

Figure 4.2 is an example of a block diagram showing a simple closed-loop control system. The diagram details a controller, and shows a plant with no detail. In the controller the command, *u*, comes in from the left, and has the feedback subtracted from it to form the error signal *e*. This error signal is applied to the k_p gain block and to an integrator through the k_i gain block, then the results are added and applied to the plant, which responds with the output to the right of the diagram.

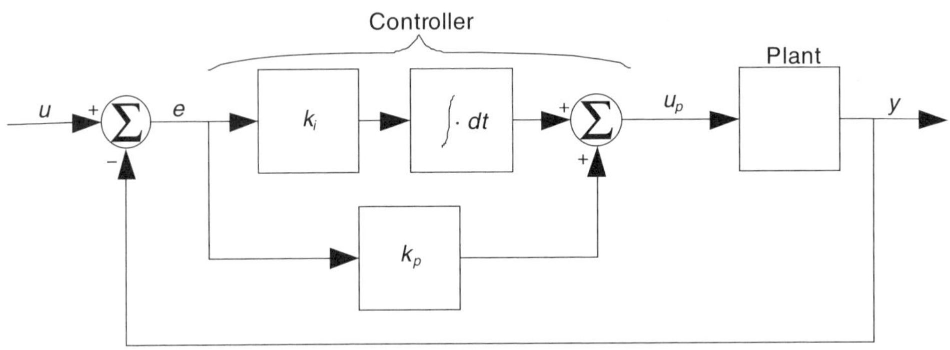

Figure 4.2 An example block diagram

It is best to keep in mind that there are no universally agreed-upon rules of block diagram structure. The language of block diagrams is a living one, and as such it evolves over time and has its own dialects. There does appear to be a consensus when one is representing systems entirely in the Laplace or z domains, but on the whole block diagramming is an imprecise science. As such, it is the responsibility of the system engineer to craft a block diagram that communicates the intended aspects of the system, and to accompany it with enough text to clarify its less common aspects.

Block diagrams consist of three main types of elements: signals, unary operator blocks, such as the "Thing" block in Figure 4.1, and m-ary operator blocks such as the two summation blocks in Figure 4.2. In addition, there are other functional elements such as sample-and-hold blocks, complex hierarchical blocks, signal sources, and so on, that don't fit in the three main categories, yet still need to be expressed if the block diagram is to be complete. Finally, it is often a good idea to annotate one's block diagram by labeling blocks, indicating areas of different sampling rates or indicating what hardware implements what functions, and so forth.

Every block has, at minimum, one signal coming out of it. In general, a block will have one or more input signals and one output signal, but blocks can be easily defined with more than one output as well.

Block diagrams can be hierarchical. Indeed, the entire block diagram detailed in Figure 4.2 could easily be the block labeled "Thing" in Figure 4.1. Furthermore, the plant block in Figure 4.2 could (and in a practical application probably would) be expanded into a block diagram of its own on a separate diagram.

Types of Elements

Signals

A signal is represented by a line with an arrowhead at the end to denote the signal direction, as shown in Figure 4.3. At your discretion, it can also be named (the signal below is '*u*'). A signal can, in general, be either a scalar signal that carries just one quantity, or a vector that contains a number of individual signals. The choice of whether to bundle signals together into a vector is up to you. To avoid confusion, the elements of a vector signal should be closely related and dissimilar signals should be kept separate.

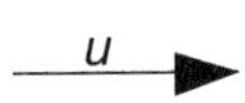

Figure 4.3 A signal

Unary Operator Blocks

A unary operator block is represented as a rectangle with signals going into and out of it, and with a symbol or some text to indicate its function. Figure 4.4 is an example of a unary operator block that operates on the input signal with the transfer function H(z) to create the output signal. Unary operator blocks are almost invariably shown with the signals traveling horizontally, either right to left or left to right. As with signals, a unary operator block can be named by placing a name above or below it.

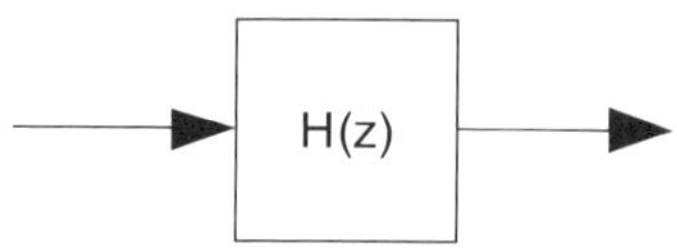

Figure 4.4 A unary operator block

A unary operator block indicates that the output signal at any given time is entirely a function of the input signal. Unary blocks can be memoryless, where the output signal at any given time is a function of the input signal at that moment. Unary blocks can also have memory, where the output signal at any given time is a function of the history of the input signal,[1] as is the case when the block implements a transfer function.

M-ary Operator Blocks

An m-ary operator block is represented as a circle, often containing a symbol or graphics to indicate its function. It will have two or more input signals and an output signal. The output signal is *always* a memoryless combination of the input signals. Figure 4.5 shows a binary summation block, where $z = x - y$.

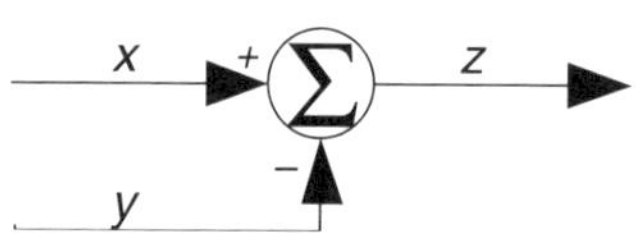

Figure 4.5 A binary summation block

1 Generally this will be the *past* history of the input signal, but occasionally an author will show a block diagram that ignores causality and shows blocks whose output signal is a function of the entire history both past and future of the input signal.

Hierarchical Blocks

A hierarchical block is simply a single block that denotes some more complex system that you show in detail somewhere else. It simply indicates that there is a system inside the block, with as many inputs as are indicated and as many outputs.

A hierarchical block can look much like a unary operator block, and when it does so can function as one in a larger block diagram. Hierarchical blocks can also indicate more general functions: they can have multiple input signals, multiple output signals, and they can be shown with "auxiliary" I/O where there is some input to or output from the block that is depreciated or designated as being fundamentally different from the "main" signals.

Figure 4.6 shows some hierarchical blocks. The leftmost one shows how you might reduce a large single input, single output subsystem down to one block to show it in a larger block diagram. The center block shows how you might represent a system with multiple inputs and multiple outputs, while the last block shows a system with an "auxiliary" input, such a mode input or an on/off input.

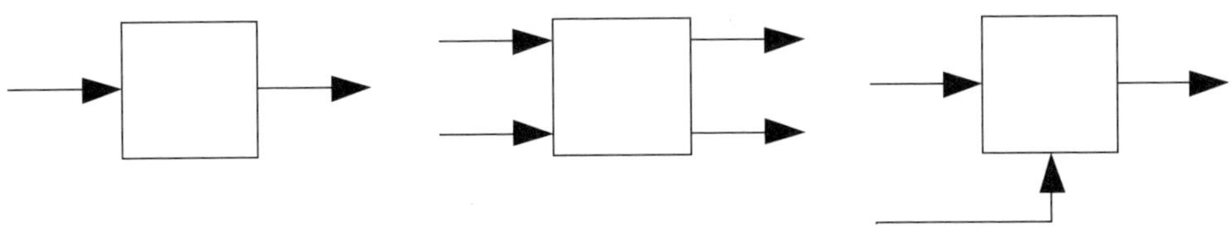

Figure 4.6 Examples of hierarchical blocks

A Dictionary of Blocks

Table 4.1 lists a number of blocks along with the function that they denote. It is by no means comprehensive, but gives a notion of the kind of things that blocks can represent.

Block	Function	Comment	
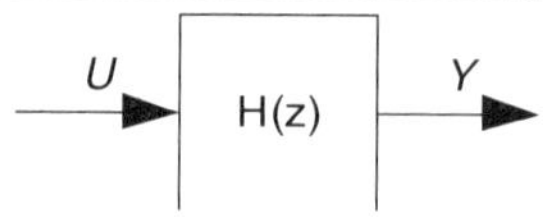	Implements the transfer function H(z): Y(z)=H(z)U(z).	Generic form for a transfer function.	(4.1)
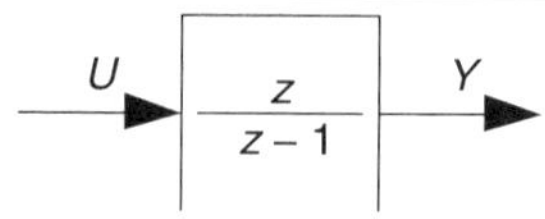	Implements a discrete integrator: $Y(z)=\frac{z}{z-1}U(z)$.	This is (4.1) with $H(z)=\frac{z}{z-1}$.	(4.2)
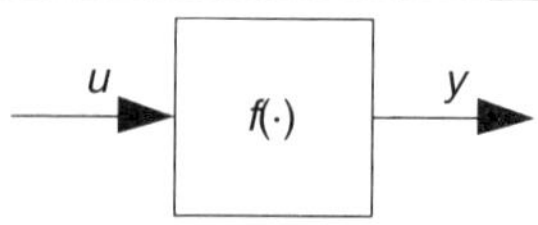	Implements the function *f*: y(t)=f(u(t)).	The function should be memoryless; this is how you show system non-linearities.	(4.3)
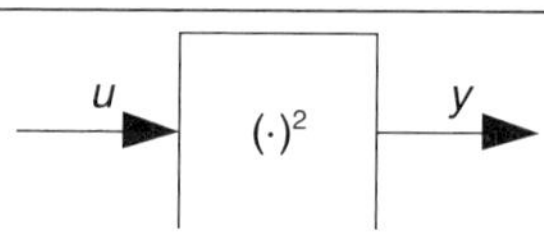	Implements the function $y(t)=u(t)^2$	A specific case of (4.3).	(4.4)
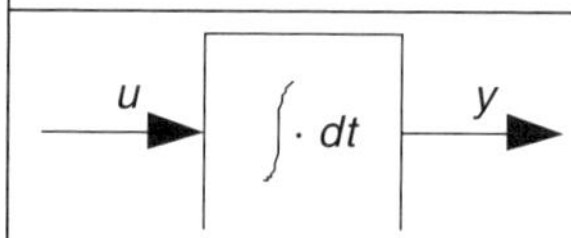	$y(t)=\int_0^t u(\tau)d\tau$	A way to show memory in the time domain.	(4.5)
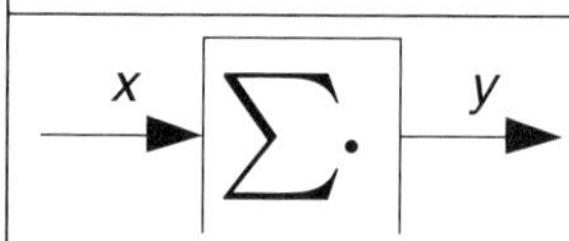	$y(n)=\sum_{k=0}^{n} u(k)$	Sampled-time version of (4.5).	(4.6)

Table 4.1 Some blocks and their meanings (continued on next page)

Block	Function	Comment	
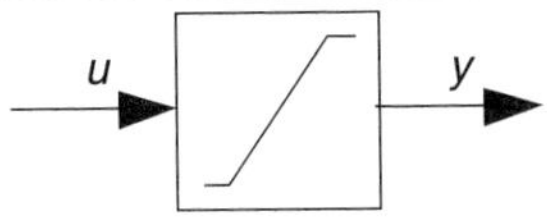	$y(t)=\begin{cases} y_{max} & u(t)\geq y_{max} \\ u(t) & y_{min}<u(t)<y_{max} \\ y_{min} & u(t)\leq y_{min} \end{cases}$	A graphical way of showing (4.3), in this case a limiter.	(4.7)
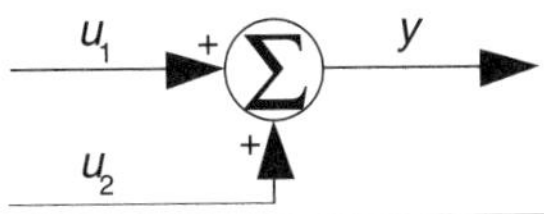	$y(t)=u_1(t)+u_2(t)$	Simple summation.	(4.8)
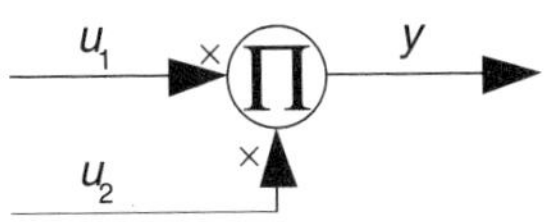	$y(t)=u_1(t)u_2(t)$	Multiplication.	(4.9)
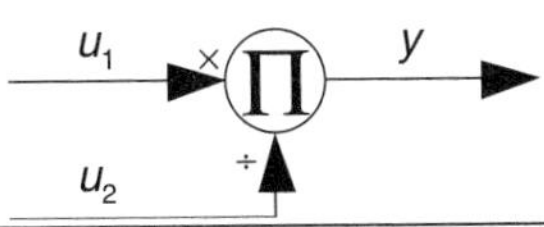	$y(t)=\dfrac{u_1(t)}{u_2(t)}$	Division.	(4.10)
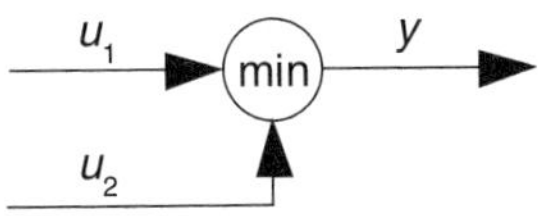	$y(t)=\min(u_1(t),u_2(t))$	Pick the minimum of the two signals.	(4.11)
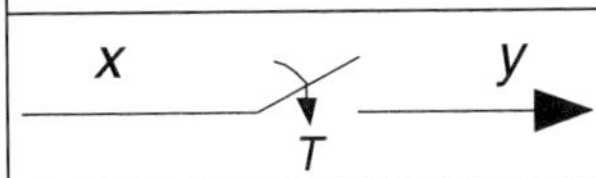	$y(n)=x(nT_s)$	Sample, see Figure 4.7.	(4.12)
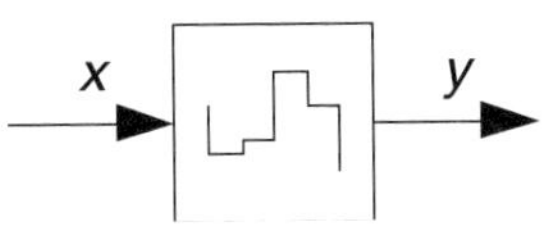	$y(t)=x\left(\text{floor}\left(\dfrac{t}{T_s}\right)\right)$	Zero-order hold, see Figure 4.8.	(4.13)

Table 4.1 Some blocks and their meanings (continued from previous page)

Sample and Hold

A sample block is denoted by a switch with a descending arrow, as shown in Figure 4.7. The period or frequency of the sample should be denoted underneath the arrowhead, or a 'T' inserted to denote a generic sample-and-hold. The function of a sample block is to sample a signal at an even rate. It is generally used to show continuous-time signals being converted to discrete-time signals for processing in a digital control system. It can also indicate a conversion from one discrete-time domain to another in systems where more than one sample rate prevails.

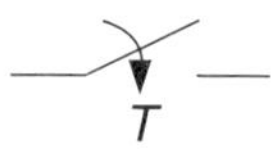

Figure 4.7 Sample

A zero-order hold is a device that transforms a discrete-time signal into a continuous-time one. Figure 4.8 shows a zero-order hold, with the graphic in the block reflecting the "staircase" appearance of a signal that's been sampled and reconstructed.

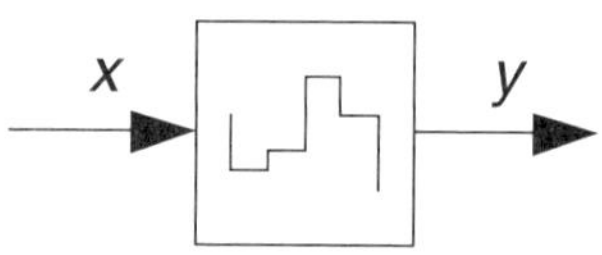

Figure 4.8 Zero-order hold

Block Diagram Dialects

Because there are no universal formal definitions of the block diagram "language," you have the freedom to invent your own dialect of the language to suit your own purposes. However, this places the responsibility for writing a clear diagram on your shoulders.

Indeed, the exact set of blocks presented here are the invention of the author; they generally follow what I have seen to be common usage, but I have made changes from common usage where I feel that my way is more clear or universal.

The common control system block diagramming language would show a summation block as in the center or right drawings in Figure 4.9; the center figure is the older form, but you will see both. I prefer the block on the left, because it shows explicitly (by means of the sigma) that a summation operation is being performed.

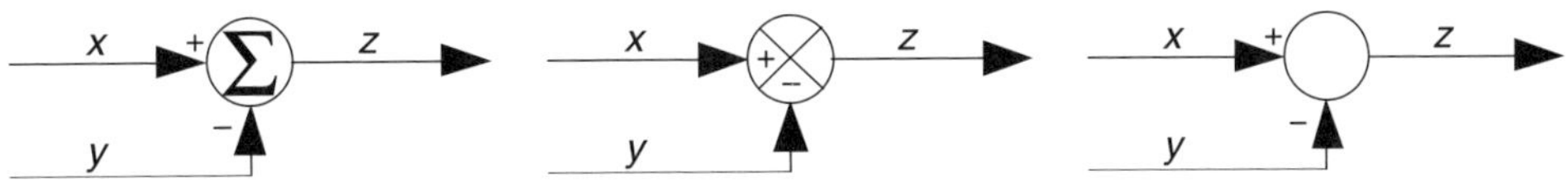

Figure 4.9 Alternate summation block

In radio usage, one will often see a frequency mixer block denoted as a circle with an 'x' inside, similar to the center summation block in Figure 4.9, yet a frequency mixer is generally a type of multiplication, which is certainly a different function from addition! Figure 4.10 shows an example, with a mixer/multiplication block on the right and an unambiguous multiplication block on the left.[2]

Figure 4.10 Multiplication or mixer block

[2] If you happen to be designing radio equipment you may still have occasion to use the mixer block on the left of Figure 4.10: mixer operation can be complex, so denoting one as a simple multiply may not be accurate—the traditional form may be less misleading.

4.2 Analyzing Systems with Block Diagrams

Block diagrams would be quite useful enough if all they were used for were to communicate system structure. They can, however, also be used to analyze system behavior, under certain sets of conditions.

If you have a system that can be represented entirely in the z domain (or Laplace), and the transfer functions of all the blocks are known, then the overall transfer function of the system can be determined from the transfer functions of the individual blocks and their interconnections.

Note that the above rule is not always altogether clear, because it is common practice to show a block diagram with a mix of z- or Laplace-domain blocks and nonlinear blocks, or with a mix of Laplace-domain blocks (indicating a continuous-time system) and z-domain blocks. Such systems can have their differential equations extracted, but they cannot be directly analyzed with the methods shown here.

Direct Equation Extraction

The most obvious method of analyzing a block diagram is to extract the relevant equations from it by inspection, then solve them directly. If the block diagram is fully specified, this technique will always deliver a mathematical model of the system.[3] As we'll see, this is not always the best way to proceed, but sometimes it is the only way.

Figure 4.11 is a quite simplified[4] block diagram of a heater control system. The plant is modeled as a heating coil that is driven by a controlled voltage whose power output will be proportional to the square of the drive voltage. The temperature of the plant is modeled as a low-pass filtered version of the ambient temperature plus a temperature difference that's being driven by the heater coil. The controller is a simple PI controller.

[3] Maybe not a solvable one, but a model none the less.

[4] It ignores the fact that the plant is continuous time, among other things.

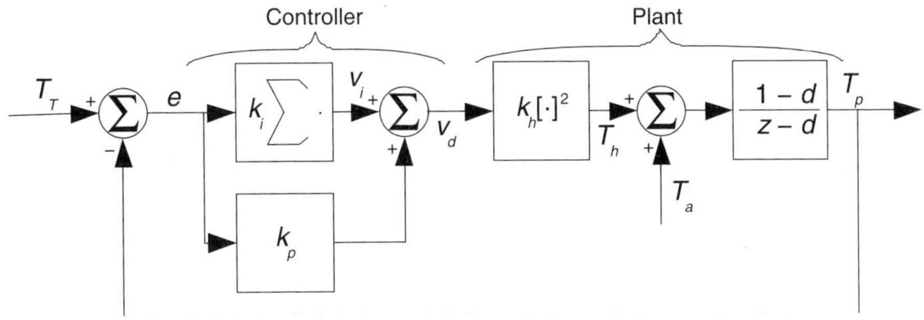

Figure 4.11 An example block diagram

Developing the equations for the diagram in Figure 4.11 is tedious but straightforward. First, find the integrator voltage:

$$v_{i,n} = v_{i,n-1} + k_i \left(T_{T,n} - T_{p,n} \right), \tag{4.14}$$

then the heater offset temperature:

$$T_{h,n} = k_h \left[v_{i,n} + k_p \left(T_{T,n} - T_{p,n} \right) \right]^2, \tag{4.15}$$

then the plant temperature:

$$T_{p,n} = T_{p,n-1} + (1-d)\left(T_{a,n-1} + k_h \left[v_{i,n-1} + k_p \left(T_{T,n-1} - T_{p,n-1} \right) \right]^2 - T_{p,n-1} \right). \tag{4.16}$$

Since the system of equations given in (4.14) and (4.16) are nonlinear they are not amenable to easy reduction, so in practice one would solve this equation by time-stepping, or by linearizing (4.16) around some operating point and solving the resulting linear difference equation (see Chapter 8 for more techniques to use in nonlinear control problems).

Manipulating Block Diagrams

Initially, one usually draws a structural block diagram. This is a diagram that shows how a system is put together. At some point, one will wish to reduce this structural block diagram into a behavioral diagram. While this can be done by the techniques shown in section 4.2 above, such techniques immediately sever the connection between the block diagram and the behavioral model, and can be very counter-intuitive to use. It is often better to reduce a block diagram using the manipulation rules presented here.

There are four tools that you have on hand to manipulate block diagrams. Given a block diagram that is described fully in the z domain or the Laplace domain, these tools will allow you to fully analyze the block diagram to extract the overall system behavior. If you observe their limitations, you can also use these tools on a variety of other block diagrams. The four tools that you have are: cascading gain blocks, moving summing junctions, combining summing junctions, and reducing loops.

The block diagram manipulations shown here will always work if the blocks in question contain pure transfer functions. All of the examples below show block manipulations on blocks containing transfer functions in the z domain; however, these manipulations can be carried out on blocks in the Laplace domain as well.

Stating when a block operation *cannot* be carried out is more difficult. In general, these operations depend on superposition and cannot be performed when the blocks contain nonlinear operation. In addition, time varying operations like sampling and zero-order holds can sometimes be moved around and sometimes cannot—generally if you can review this section and make the mathematical equations fit, then you can perform the operation with a block diagram.

Loop Reduction

When a block diagram indicates a feedback loop, you can reduce the loop to a single transfer function block as shown in Figure 4.12. If you look at the equations that govern the behavior of this block diagram, you can see that the output signal is a function of the forward gain G and the error signal *e*:

$$Y(z) = E(z)G(z). \tag{4.17}$$

The error signal, in turn, is a function of the input and output signals and the feedback gain:

$$E(z) = U(z) - Y(z)H(z). \tag{4.18}$$

By substituting expressions, we get

$$Y(z) = \left[U(z) - Y(z)H(z)\right]G(z), \tag{4.19}$$

which reduces to

$$\frac{Y(z)}{U(z)} = \frac{G(z)}{1 - H(z)G(z)}. \tag{4.20}$$

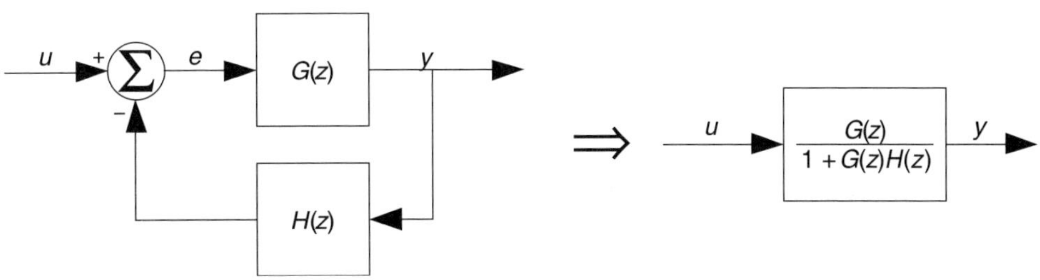

Figure 4.12 Reducing a loop

Example 4.1 Using Loop Reduction

A feedback control system has a forward gain

$$G(z) = \frac{0.4z}{z^2 - 1.64z + 0.64} \tag{4.21}$$

and negative feedback with a gain of

$$H(z) = 0.1. \tag{4.22}$$

Draw its block diagram, and find the overall transfer function for the system.

The block diagram is a simple feedback loop with the specified gains:

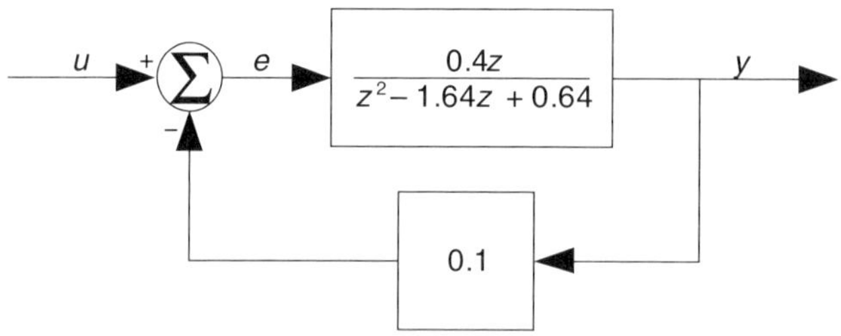

Figure 4.13 Feedback loop for Example 4.1

From the formula for loop reduction, the transfer function for the system is

$$\frac{Y(z)}{U(z)} = \frac{0.4z}{z^2 - 1.6z + 0.64}. \tag{4.23}$$

This reduces down to the block diagram:

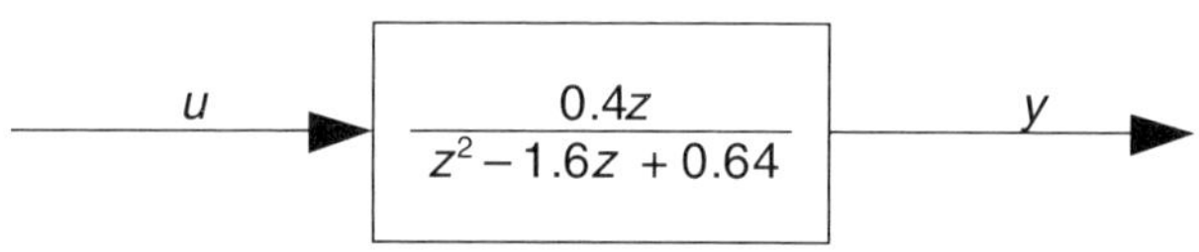

Cascading Gains

When two blocks are cascaded directly, the transfer function of the combination is the product of the two transfer functions, as in Figure 4.14. Looking at the equations that govern the behavior of the block diagram you can see that

$$X(z) = U(z)G_1(z) \text{ and } Y(z) = X(z)G_2(z). \tag{4.24}$$

From this, it is easy to see that

$$Y(z) = U(z)G_1(z)G_2(z). \tag{4.25}$$

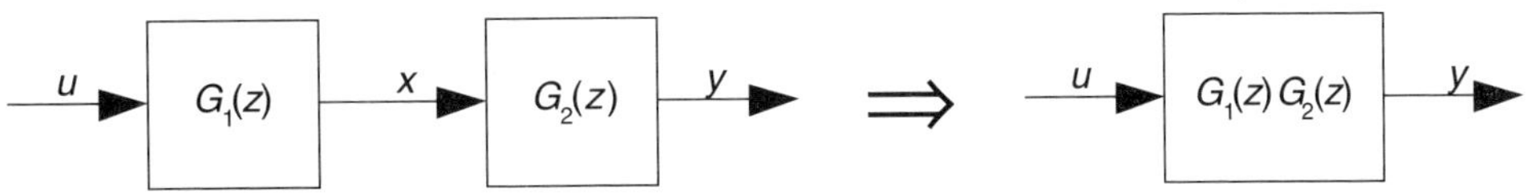

Figure 4.14 Cascading gain blocks

Example 4.2 Cascading Gains

A feedback control system has a plant with a gain of

$$G_2(z) = \frac{z}{z^2 - 2z + 1}, \tag{4.26}$$

a controller with a gain of

$$G_1(z) = \frac{0.584z - 0.268}{z - 0.384}. \tag{4.27}$$

and unity feedback. Draw its block diagram, and find the overall transfer function for the system.

The block diagram is:

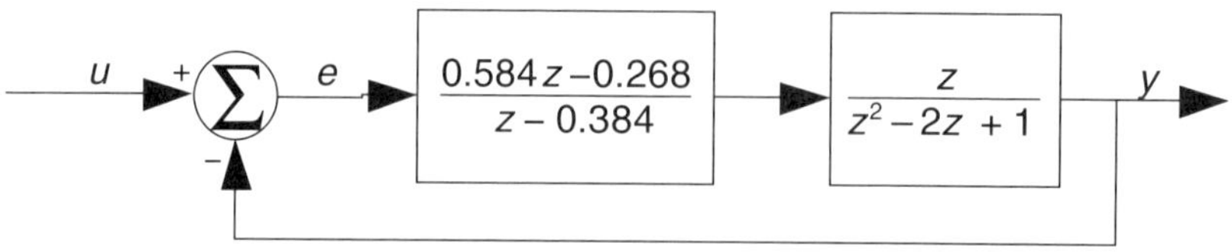

From the gain cascade rule the forward gain is

$$G(z) = \frac{0.584z - 0.268}{z - 0.384} \cdot \frac{z}{z^2 - 2z + 1},$$

$$G(z) = \frac{0.584z^2 - 0.268z}{z^3 - 2.384z^2 + 1.768z - 0.384}. \tag{4.28}$$

Using the loop reduction rule with $H(z) = 1$, the transfer function for the system is

$$\frac{Y(z)}{U(z)} = \frac{0.584z^2 - 0.268z}{z^3 - 2.2z^2 + 1.5z - 0.384}. \tag{4.29}$$

This reduces down to the block diagram:

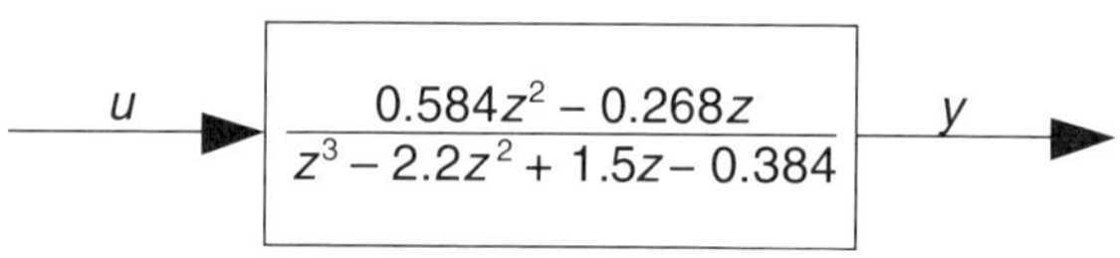

Summing Junctions

If a loop contains more than one summing junction, it cannot be reduced by simple loop reduction, and cascading gains will not eliminate the extra junction. In order to reduce such a loop, the summing junctions must be moved around until there is a loop with just one summing junction. This is done by propagating a transfer function backward through the junction, or its inverse forward through the junction, as shown in Figure 4.15.

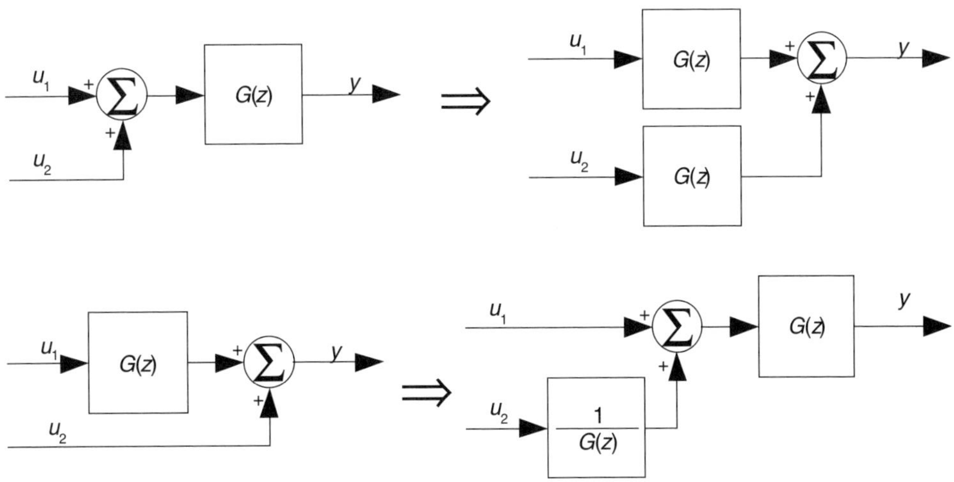

Figure 4.15 Moving summing junctions

Keep in mind that the inverse transfer function $1/G(z)$ may not be physically realizable—this is not a concern unless you are trying to use your intermediate results in a real system. Any manipulations you do here are just for the purposes of making the math easy, and if the system starts as a physically realizable one then any contradictions will disappear by the time you get your solution.

In the top case of Figure 14.5, the input/output relationship on the left is

$$Y(z) = [U_1(z) + U_2(z)]G(z). \tag{4.30}$$

Using the commutative property of multiplication, this translates to

$$Y(z) = U_1(z)G(z) + U_2(z)G(z), \tag{4.31}$$

which corresponds to the input/output relationship on the right.

In the bottom case of Figure 4.15 the input/output relationship on the left is

$$Y(z) = U_1(z)G(z) + U_2(z). \tag{4.32}$$

By using the inverse of the transfer function, the right side becomes

$$Y(z) = \left[U_1(z) + \frac{U_2(z)}{G(z)}\right]G(z), \tag{4.33}$$

which is functionally the same as the left.

When a block diagram contains a pair of signal paths that originate from the same signal and terminate on a summing junction, the set can be reduced to a single block, as shown in Figure 4.16. In this case it can be seen that

$$Y(z) = U(z)G_1(z) + U(z)G_2(z). \tag{4.34}$$

From this it is easy to get

$$Y(z) = U(z)\left[G_1(z) + G_2(z)\right]. \tag{4.35}$$

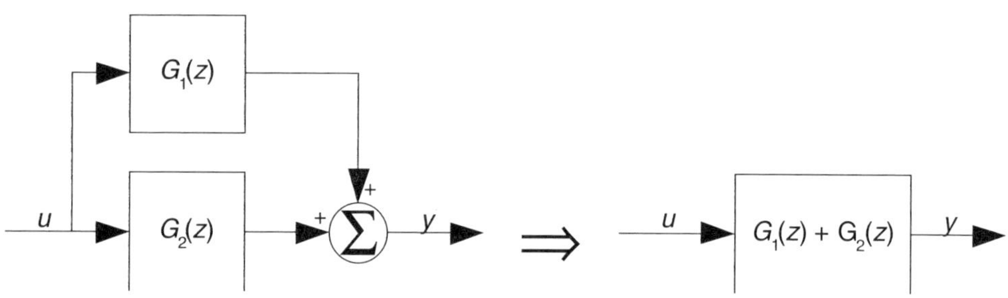

Figure 4.16 Removing parallel paths

Example 4.3 Cascading Gains

Feedforward is often used in control systems to increase the system response speed without having to change a "safe" set of tuning parameters for the loop. Figure 4.17 shows one such system. We can reduce the block diagram in Figure 4.17 down to a single block.

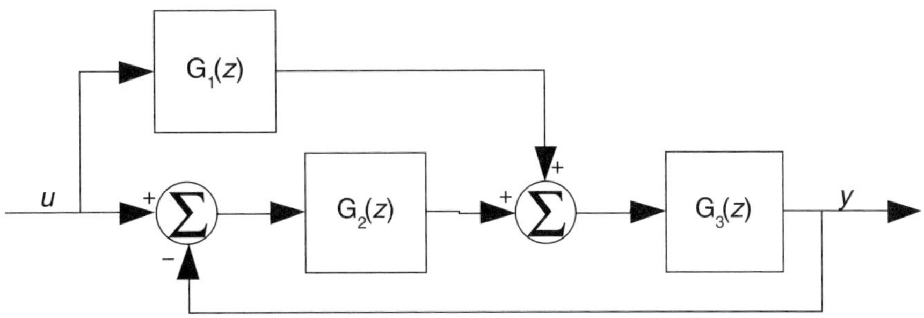

Figure 4.17 A system with feedforward

The loop in Figure 4.17 contains two summing junctions, which prevents the use of the loop reduction rule. We can either move the G_1 leg back to the left summing junction, cascading G_1 with $1/G_2$, or we can move the feedback leg forward to the right summing junction, putting a G_2 block in the feedback path. I'll do the latter, to avoid the inverse:

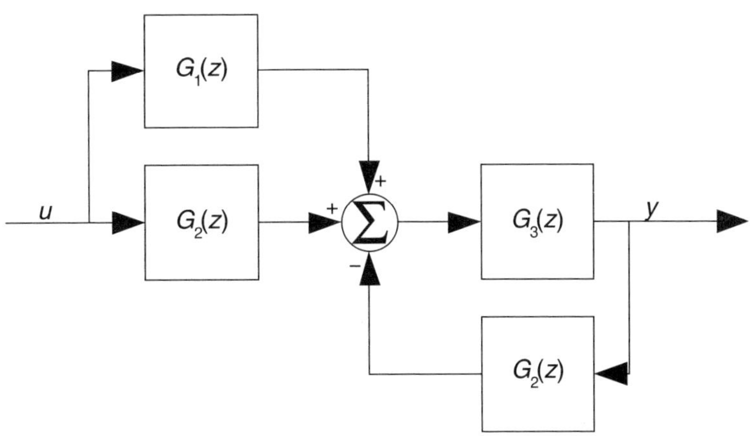

Using loop reduction for the right half and the parallel summation rule for the left gives us a transfer function of:

$$Y(z) = U(z)\frac{\left[G_1(z) + G_2(z)\right]G_3(z)}{1 + G_2(z)G_3(z)}. \tag{4.36}$$

Multiple Input Systems

So far I've presented systems with just one input and one output (often called SISO systems for "Single Input, Single Output"). Real systems aren't restricted to having just one input and one output, and neither are block diagrams. Even when you're working with a system that you'd like to treat as a SISO system, you'll find that reality may place additional requirements on your analysis.

It is often very useful to model a disturbance to a system in a block diagram. Figure 4.18 shows a block diagram[5] with the plant split up into an actuator and a mechanism, and a disturbance force added to the force of the actuator on the mechanism. This is an example of a multiple input, single output (MISO) system. When dealing with such a system, you are often interested in its ability to reject disturbances, so you often want to find the transfer function from the intended input (u in this case) as well as the disturbance input (u_d).

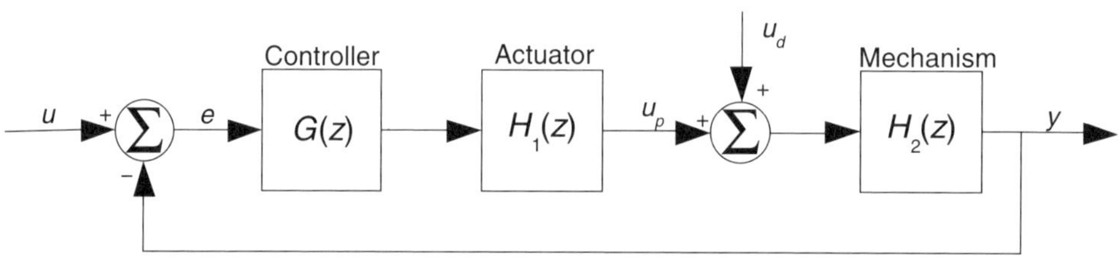

Figure 4.18 A block diagram showing a disturbance input

[5] Ignoring the continuous-time nature of the actuator and plant.

Both of the transfer functions in question can be found almost by inspection from Figure 4.18. This is done by appealing to the fact that we're using a linear system model, which means that we can use superposition. To find the transfer function from the intended input to the output, we assume that the disturbance is set to zero and solve the system normally:

$$T_i(z) = \frac{Y(z)}{U(z)} = \frac{G(z)H_1(z)H_2(z)}{1+G(z)H_1(z)H_2(z)}. \tag{4.37}$$

Similarly, to find the transfer function from the disturbance input to the output, we set the intended input to zero and solve the resulting system:

$$T_d(z) = \frac{Y(z)}{U_d(z)} = \frac{H_2(z)}{1+G(z)H_1(z)H_2(z)}. \tag{4.38}$$

Notice that the denominators in (4.37) and (4.38) are the same. This is a general characteristic of feedback control systems: no matter how many inputs or outputs the system has, no matter how the system behavior may vary when you choose different points to inject signals or observe them, the underlying behavior of the system in terms of the poles of the transfer function will remain the same. The only times that you will see two different characteristic polynomials[6] in one system will either be because you have two disjoint systems that happen to be on the same page, or because the numerator and denominator of the system share some roots.[7]

There is one other thing to notice about the system: it is often not necessary to derive all of the transfer functions directly. In this case one could have observed that by moving the summing junction for the disturbance back to the input summing junction the disturbance transfer function could be found from (4.37):

$$T_d(z) = \frac{1}{G(z)H_1(z)}T_i(z). \tag{4.39}$$

---

[6] See Chapter 2.

[7] This is called *pole-zero cancellation*. The cancelled poles haven't ceased to exist—they are just hidden, often to the detriment of real system behavior. Systems with pole-zero cancellation should be treated with all due caution.

If you do the math, you'll see that (4.39) and (4.38) are equal. In many cases it is much easier to derive the desired function by this method rather than solving the block diagram twice.

Example 4.4 A Heater with PI Control

Figure 4.19 shows a temperature control system where a plant's temperature is kept constant by controlling the power applied to a heater, but is disturbed by the ambient temperature (u_d) The plant transfer function is given the rather simplistic model[8]

$$H(z) = \frac{(1-d_1)(1-d_2)}{(z-d_1)(z-d_2)}. \tag{4.40}$$

The controller is a PI, with transfer function

$$G(z) = k_p + \frac{k_i}{z-1}, \tag{4.41}$$

where the controller gains have been adjusted for a stable, rapidly settling plant.

Find the system's response to a change in ambient temperature and solve for the steady-state component of this response.

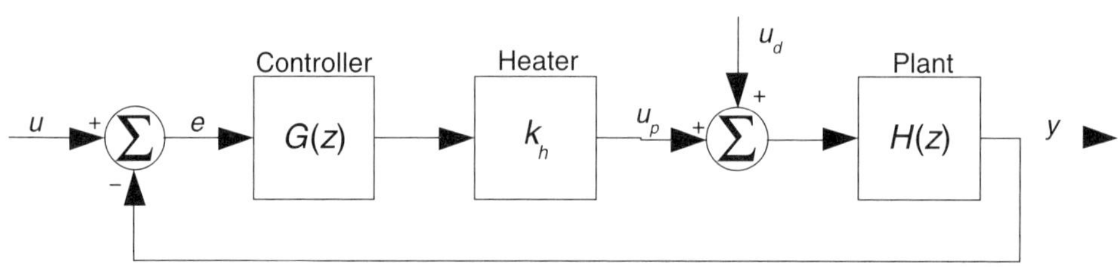

Figure 4.19 A heater control system affected by ambient temperature

---

8 Thermal systems tend to have very complex dynamics, which are difficult to model with traditional z transforms. One is often reduced to using measured frequency response data and the frequency-domain design techniques, presented in Chapter 5, without ever seeing an explicit transfer function.

Borrowing from (4.38), the disturbance transfer function is

$$T_d(z) = \frac{Y(z)}{U_d(z)} = \frac{H(z)}{1 + G(z)k_h H(z)}. \tag{4.42}$$

Substituting in (4.40) and (4.41), this becomes

$$T_d(z) = \frac{Y(z)}{U_d(z)} = \frac{\dfrac{(1-d_1)(1-d_2)}{(z-d_1)(z-d_2)}}{1 + k_h\left(k_p + \dfrac{k_i}{z-1}\right)\dfrac{(1-d_1)(1-d_2)}{(z-d_1)(z-d_2)}}. \tag{4.43}$$

Let $k_{pp} = (1-d_1)(1-d_2)$ and simplify to find the response to a change in ambient temperature:

$$\begin{aligned} T_d(z) &= \frac{Y(z)}{U_d(z)} = \frac{k_{pp}(z-1)}{(z-d_1)(z-d_2)(z-1) + k_h(k_p z + k_i - k_p)k_{pp}}, \\ T_d(z) &= \frac{k_{pp}(z-1)}{z^3 - (d_1+d_2)z^2 + (d_1 d_2 + d_1 + d_2 + k_h k_{pp} k_p)z + k_h k_{pp}(k_i - k_p) - d_1 d_2}. \end{aligned} \tag{4.44}$$

To find the response to a steady state change in ambient temperature it is necessary to make a couple of observations: First, a steady-state change in ambient temperature will have a number of components that go to zero as time goes to infinity, plus a unit step response. Second, we specified that the PI controller would be tuned such that the system is stable. This ensures that all of the system poles lie in the set of z such that $|z| < 1$.

With the first of these two observations, we can apply the final value theorem from Chapter 2 to get an expression for the steady state error:

$$\text{error}_{\text{steadystate}} = \lim_{z \to 1} (z-1)\left(\frac{z}{z-1}\right)T_d(z). \tag{4.45}$$

With the second of these two observations, we can deduce that the denominator of $T_d(z)$ will be a finite number as z goes to 1; thus (4.45) can be simplified to

$$\text{error}_{\text{steadystate}} = \lim_{z \to 1} (z-1)\left(\frac{z}{z-1}\right)\frac{k_{pp}(z-1)}{k_0}. \tag{4.46}$$

where k_0 is known to be nonzero. At this point, it is easy to see that the steady-state error is equal to zero.

Multiple Output Systems

A dual of the case with multiple input, single output (MISO) systems are systems with single inputs and multiple outputs (SIMO). As with the multiple input case, one often finds the multiple output case when we would rather be doing a simpler analysis.

Take the case of the command-following system in Figure 4.20 (which is just Figure 4.18 with the extra summing block removed and some extra signal labels). With such systems, we're often concerned not only with how well the output is going to follow the input, but what size the control signals will be for a given input.

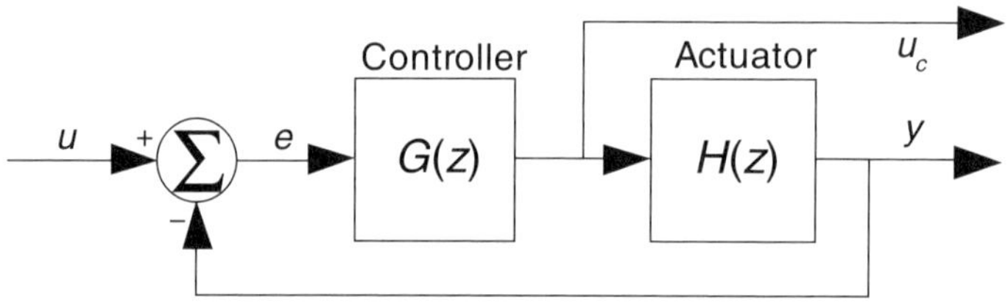

Figure 4.20 A command-following system

If we have a known input and we wish to know the drive signal, we only need to find the transfer function from input to drive and apply the input signal. In this case the block diagram can be reduced by inspection:

$$\frac{U_c}{U} = \frac{G(z)}{1 + G(z)H(z)}. \tag{4.47}$$

This transfer function can be used to predict the drive signal to the actuator for the purposes of checking on actuator heating, actuator drive requirements, and to see if any system parameters are being exceeded.

Example 4.5 Positioning System Drive Requirements

A system as shown in Figure 4.20 is built with a voltage-driven, geared motor with a time constant of approximately 20ms. The control system samples at 100Hz, so the effective transfer function of the motor is

$$H(z) = \frac{Y(z)}{U_c(z)} = \frac{0.00426z + 0.00361}{(z - 0.607)(z - 1)} \tag{4.48}$$

for per-unit drive and a motor output in radians. The controller has the transfer function

$$G(z) = \frac{9.5z - 5}{z - 0.4}. \tag{4.49}$$

Find the transfer function from the system command input to the motor drive input. Find the largest step command that can be given to the system before the motor's per-unit drive exceeds an absolute value of 1. Find the maximum ramp command that can be given to the system without the motor's per-unit drive exceeding an absolute value of 1. Find a way that the motor could start from a stop, be commanded to move over 50% of its range, then stop without exceeding the maximum motor drive.

From (4.47) the transfer function from the command to the motor drive is

$$\frac{U_c}{U}=\frac{(9.5z-5)(z-0.607)(z-1)}{(z-0.607)(z-1)(z-0.4)+(0.00426z+0.00361)(9.5z-5)} \tag{4.50}$$

or

$$\frac{U_c}{U}=\frac{9.5z^3-20.27z^2+13.80z-3.035}{z^3-1.9727z+1.2722z-0.2641}. \tag{4.51}$$

Then the response to a unit step is found by multiplying (4.51) by a step:

$$U_{cstep}=\frac{z}{z-1}\frac{9.5z^3-20.27z^2+13.80z-3.035}{z^3-1.9727z+1.2722z-0.2641}. \tag{4.52}$$

$$U_{cstep}=9.5+\frac{7.40z-4.87}{z^2-1.520z+0.584}+\frac{0.518}{z-0.447}. \tag{4.53}$$

This response can be seen in Figure 4.21. As shown in the figure, the maximum value of the response occurs at the initial timestep, so the height of the response can be found by applying the initial value theorem to (4.52), which gives a value of 9.5. Thus, the largest step command that can be given without exceeding a drive of 1 is equal to 1/9.5 = 0.105.

The motor drive response to a unit ramp input can be found in a similar manner. Multiplying (4.51) by a ramp gives

$$U_{cramp}=\frac{z}{(z-1)^2}\frac{9.5z^3-20.27z^2+13.80z-3.035}{z^3-1.9727z+1.2722z-0.2641}. \tag{4.54}$$

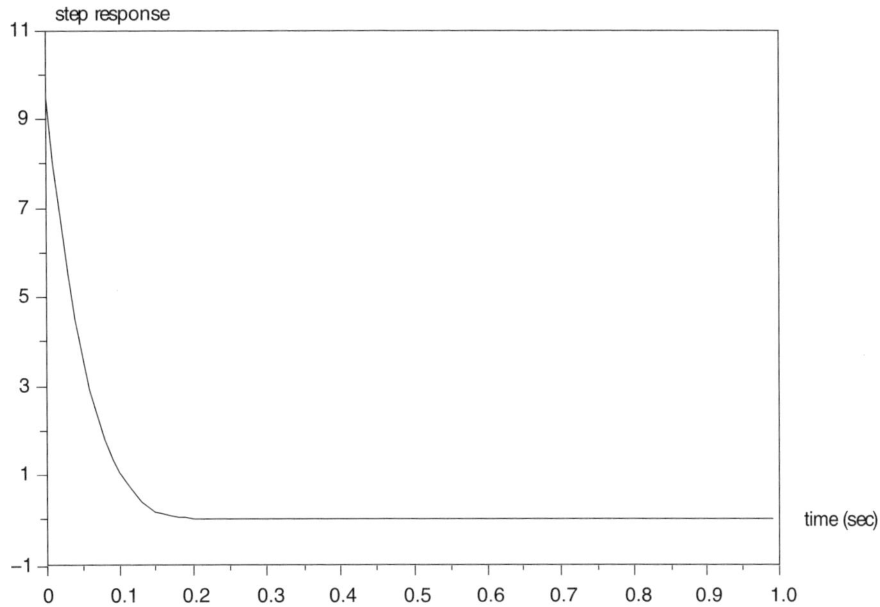

Figure 4.21 Drive to motor from a unit step input

Figure 4.22 shows the time-domain response. Clearly, the final value of this response is also its maximum, so the maximum motor drive can be found using the final value theorem. The final value of U_{cramp} is:

$$U_{final} = \lim_{z \to 1} \frac{1}{z-1} \frac{9.5z^3 - 20.27z^2 + 13.80z - 3.035}{z^3 - 1.9727z + 1.2722z - 0.2641} = 49.9. \quad (4.55)$$

Thus, the maximum ramp that can be imposed is 1/49.9 per step, or approximately 2%/step, or 200%/second.

There are a number of ways that the motor can be driven to a position without exceeding its maximum drive. One such way is shown in Figure 4.23, where the rate of change of the position command is limited to a 2% per sample ramp. The resulting position command, motor drive command, and motor position response are all shown, and as can be seen the motor command stays within appropriate limits.

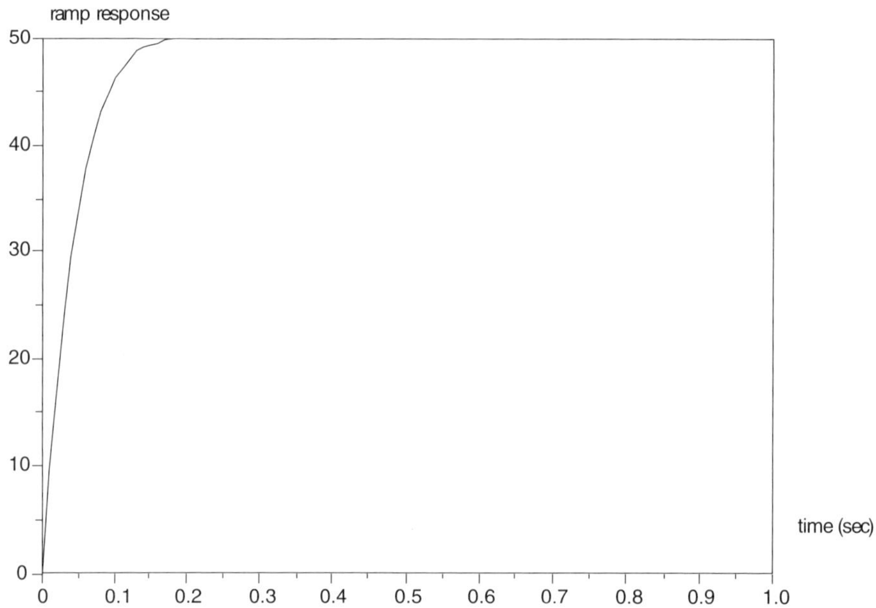

Figure 4.22 Drive to motor from a unit ramp input

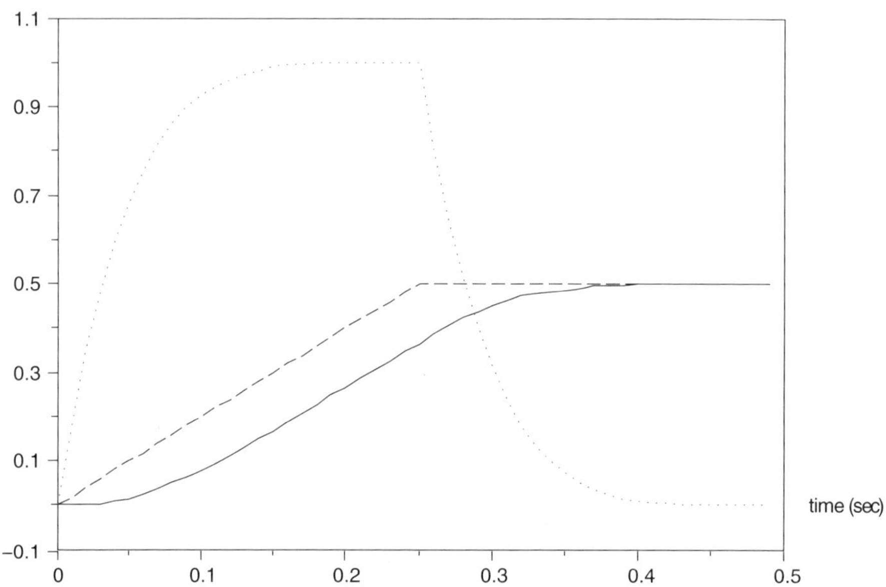

Figure 4.23 Driving motor with a ramp

4.3 Conclusion

This chapter presented a method of describing a system using block diagrams. It showed how these block diagrams may vary within the control systems community and in the larger signal processing community. I presented methods for bringing some formalism to a block diagram description so that it could be used to analyze system behavior, and methods for manipulating block diagrams to discover how a particular system will behave in a larger sense.

5

Analysis

Previous chapters have shown how to model a specific system and describe it from a strict z-transform point of view. An inescapable fact of control system design, however, is that plant models are almost never accurate. Any given plant's characteristics will change with time, temperature, or perhaps the direction of the wind. In addition, if you are designing embedded controllers for systems that are to be produced in quantity, each plant will also be different from the next, and from the "standard" plant to which you designed in the first place.

If you are to work effectively with control systems, you must be able to predict how your system will behave, not only for the specific case of the plant as you have modeled it, but for all the variations that your system will throw at you as conditions change.

A description of a static system isn't an analysis of what will happen with the system when conditions change. This chapter presents methods for analyzing how systems will behave as their characteristics change. All of these methods add value if the system can be modeled with a strict z-domain transfer function. In addition, the two frequency-domain methods presented here will work when one lacks a z-domain model, as long as you have measured frequency domain information.

In this chapter I'll show you the root locus plot, which lets you get a pictorial representation of what the poles in your system are doing. Then I'll show you a number of methods to use a system's tested response to a sine wave to do control system design, and how those relate to the root locus plot. Finally, I'll use these tools to show how a variety of controllers and techniques can be used for system design.

5.1 Root Locus

In Chapter 2 we saw that we could view the system pole and zero locations graphically. This is handy, but one can go beyond that and generate plots that show the loci of all possible roots of the system's characteristic polynomial as some parameter in the system is varied. Since such a plot shows instantly whether and at what point a system will go unstable when the parameter is varied, they can be very useful for system design. Such a plot is called a *root locus* plot, because it shows the loci of all possible roots of the system characteristic polynomial.

There are two primary uses for the root locus plot. First, you can use the plot in design, by showing yourself the effect of including different controllers in the system. With careful choice of the parameter to be varied, you can find whether or not a particular controller will get you the performance that you are looking for. Second, you can choose a parameter that reflects how your system may vary over time or over a production run, and you can get a feel for how well your system will perform in the face of these changes.

The Root Locus in General

Consider a system that is parameterized by a variable θ:

$$H(z) = \frac{\sin\theta\cos\theta z}{z^2 - (2 - \cos\theta - \sin\theta\cos\theta) z + 1 - \cos\theta}. \tag{5.1}$$

Figure 5.1 is the root locus plot of (5.1) as θ is varied from 0 to 2π. I constructed it by iterating the value of θ and computing the roots of the resulting polynomial, then plotting the results on the complex plane along with the unit circle (dotted). From this plot you can see that as the parameter is varied, the transfer function takes on both stable and unstable values. Additional digging, in fact, revealed that this transfer function is only stable for values of θ between 0 and $\pi/2$.

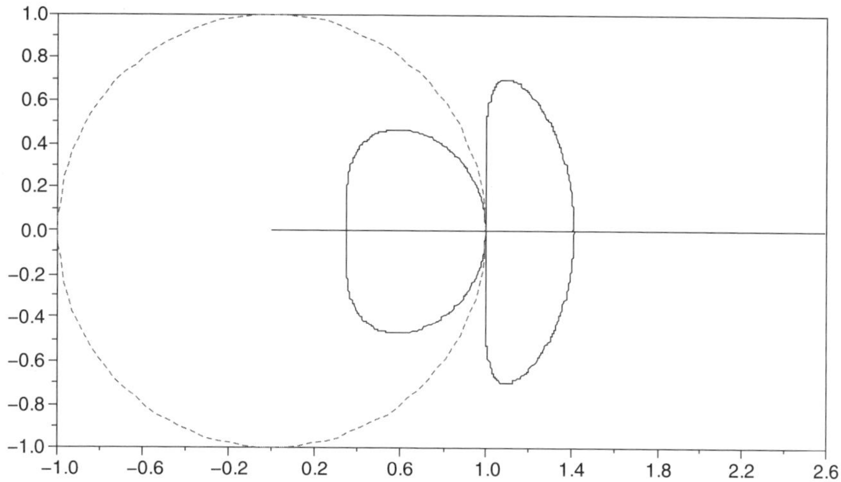

Figure 5.1 Root locus plot of (5.2)

The Evans Root Locus Plot

Someone who has been practicing control engineering might look at the example in (5.1) and exclaim "Hey! That's not the root locus plot that I learned!" They'd be right, because what is generally taught is the Evans root locus.

When a system is parameterized so that the coefficients of the system polynomial vary as a linear function of the parameter then an Evans root locus plot can be constructed for the system. The Evans root locus method restricts the type of system that can be plotted, but it has properties that make it possible to sketch a root locus plot by hand, and many problems in control systems can be reduced to an Evans root locus plot.

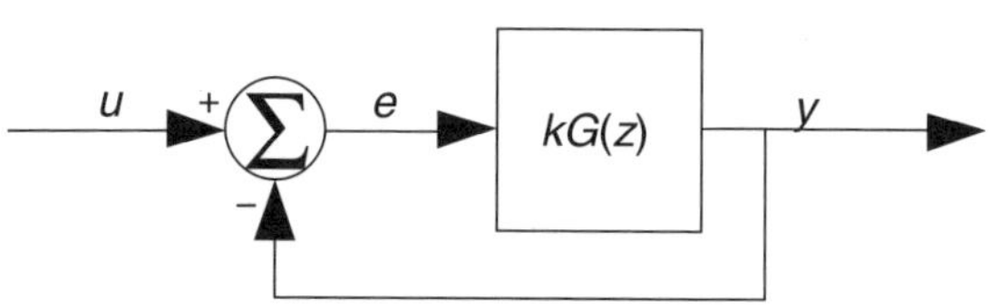

Figure 5.2 A unity-feedback control system

Consider a basic unity-feedback system such as the one shown in Figure 5.2, with a block transfer function equal to a gain, k, times a forward transfer function, G. The transfer function is:

$$H(z) = \frac{Y(z)}{U(z)} = \frac{kG(z)}{1+kG(z)}. \tag{5.2}$$

Now take G as a ratio of polynomials, so

$$G(z) = \frac{A(z)}{B(z)} \tag{5.3}$$

or

$$G(z) = \frac{a_n z^n + a_{n-1} z^{n-1} + \cdots + a_0}{z^n + b_{n-1} z^{n-1} \cdots b_0}. \tag{5.4}$$

Then one can substitute (5.3) into (5.2) to get

$$H(z) = \frac{kA(z)}{kA(z) + B(z)}. \tag{5.5}$$

Now the system characteristic polynomial has coefficients that are a constant plus a linear factor of the gain, k:

$$kA(z) + B(z) = (1 + ka_n) z^n + (b_{n-1} + ka_{n-1}) n^{n-1} + \cdots + b_0 + ka_0. \tag{5.6}$$

The root locus that results from varying k is an Evans root locus. Of course you can construct a root locus plot of the system by iterating the value of k through an appropriate interval and evaluating the roots of (5.6) at each step, as was done for the plot in Figure 5.1. In fact, with today's computers this is probably the most efficient way to do it. Several mathematics packages[1] either come with an Evans root locus plotter built in or have add-ons that will plot the Evans root locus.

1 SciLab and Matlab® (with the Control Systems Toolbox).

If your math package doesn't support the generation of the root locus, but does support polynomial factoring it is easy to write your own root locus generator.

Before the advent of the fast digital computer, however, root locus plots were constructed by hand, and because the Evans root locus method is very practical to do by hand, in the past Evans root locus plots were almost universal. There is a plethora of rules that assist in the hand-plotting of the Evans root locus, and for refining results found by these graphical methods. While it is no longer efficient to construct exact root locus plots by hand, the properties of the root locus plot still help us to understand and design control systems. Furthermore, it is often useful to be able to sketch a quick root locus plot as an aid to understanding the effect that a particular control strategy may have on one's system's behavior.

Take the example of the system in Figure 5.2, with

$$G(z) = \frac{z-0.5}{(z-1)(z-0.9)}. \tag{5.7}$$

Figure 5.3 shows the root locus plot of the system as k is varied. The two x's denote the positions of the poles of G(z), and the diamond denotes the zero. The lines show the path of the poles of the system as k is varied, with the poles

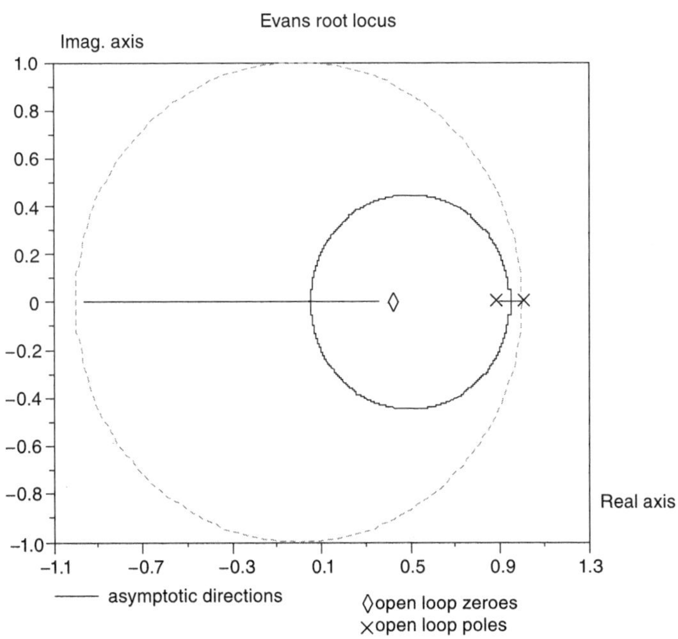

Figure 5.3 Root locus of (5.7)

starting at the open-loop poles, and ending with one closed-loop pole going toward the system zero and the other closed-loop pole heading off toward negative infinity.

Root Locus Properties

I am going to state, largely without proof, a number of properties of root locus plots. Some of these properties are specific to the Evans root locus plot, but many of them are more general; I have noted where the properties are general.

The Root Locus is Continuous

This is a property of any root locus: as long as the polynomial coefficients are varied continuously and the leading term in the polynomial, such as the z^2 term in (5.1), does not go to zero as the parameter is varied, the root locus plot is continuous, with no "jumps" from one point to another. In the Evans case, this reduces to the leading term never going to zero as k is varied.

Points on the Plot

The Evans method is based on the simple observation that if k is a positive real number then a point z_c is on the root locus if and only if $G(z_c)$ is real and strictly negative. You can arrive at this observation by the following train of reasoning:

- By definition the point z_c is on the root locus if and only if the system characteristic polynomial has a zero at z_c for some positive value of k.
- The system characteristic polynomial has a root at some z_c and k if and only if the denominator of (5.5) is zero at that z_c and k.
- The denominator of (5.5) is zero if and only if the denominator of (5.2) is zero.
- The denominator of (5.2) is zero if and only if $k\ G(z_c) = 1$.
- The expression $k\ G(z_c) = 1$ is true if and only if $G(z_c)$ is strictly negative.

A great many rules for the hand-construction of the Evans root locus plot are derived from this one rule. I won't expand on it to that extent, but it is an important part of using the root locus, and it will be very useful later on for understanding the frequency domain methods I will present.

Roots Travel to Zeros or Infinity

Another property of the Evans root locus is that as k is varied from zero to infinity the root locus starts at the open-loop poles then travels to the open-loop zeros. Any excess roots end up going of to infinity.

Any root locus whose parameters rise monotonically with the parameter being varied will share this property with the Evans root locus.

If you look at (5.5) you'll see that when $k = 0$ the denominator of $H(z)$ is just $B(z)$, so the system poles must be equal to the poles of $G(z)$. When k tends toward infinity the denominator is dominated by $A(z)$, so the poles go to the roots of $A(z)$—except for the excess poles that arise when the order of $B(z)$ is higher than the order of $A(z)$. Because any system with poles whose magnitude exceeds 1 is unstable this means that there will *always* be some maximum k that will drive the system into instability. Real systems always have at least one excess pole, so in general you can count on the system going unstable with a too-high value of k.

Using Root Locus

A root locus plot tells you all possible locations for the poles of your system's transfer function as some parameter is varied. This property is directly useful three ways: to find the correct parameters for a chosen controller, to determine if a system will remain stable as plant parameters change, and to investigate particular types of controllers for their suitability to solve the problem at hand.

Designing using the root locus involves making some controller decisions, then checking them either to refine the controller or to verify that the system will retain its desired properties as the plant varies. To find the correct controller parameter, one varies the parameters to find the best value. To see the effect of varying a plant parameter, vary the parameter through the range of interest, and verify that the system poles don't stray out of the preferred region.

One shouldn't limit root locus techniques to just checking stability—it is easy to choose some region within the stability region as your "preferred" pole locations, then use the root locus to make sure that the locus stays in that region.

Example 5.1 A Motor Controller using Root Locus

A motor driven positioner is to be controlled, with a requirement that the final system poles have absolute values no greater than 0.7 and are on the real number line.

A DC motor driven by a voltage is to be used. The positioner transfer function is

$$G_p(z) = \frac{Az}{(z-0.9)(z-1)}, \tag{5.8}$$

where A is a characteristic gain of the system. The system arrangement shown in Figure 5.4 with a PD controller has been proposed to control the positioner. The controller transfer function is

$$G_c(z) = k_p + k_d \frac{z-1}{z}. \tag{5.9}$$

Use root locus methods to determine if the controller is suitable for the job, and to find the correct values for the proportional and derivative gains.

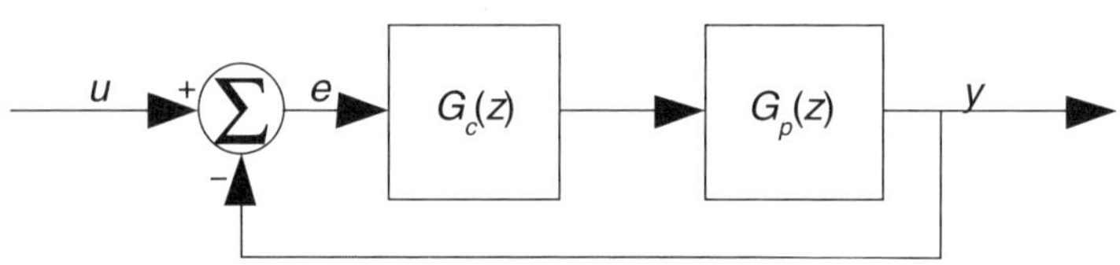

Figure 5.4 A unity-feedback control system

The open-loop transfer function for the system is

$$G_{ol}(z) = \left(k_p + k_d \frac{z-1}{z}\right)\left(\frac{Az}{(z-0.9)(z-1)}\right). \tag{5.10}$$

This is, unfortunately, not in a form that can immediately be used in an Evans root locus plot, because we have two parameters—k_p and k_d—that can be varied. But (5.10) can be easily modified to

$$G_{ol}(z) = k_p \frac{(z - d_z)A}{(z - 0.9)(z - 1)} \tag{5.11}$$

where $d_z = k_d / k_p$. If we set d_z to 0.5 then the root locus plot of Figure 5.3 results.

Unfortunately this value of d_z is not part of a suitable solution, because there are no poles with critical damping and a magnitude of 0.7! However, if we make the derivative term of the controller more aggressive it will tend to increase d_z, and this will tend to make the circular portion of the root locus smaller. After some experimentation it is found that a value of d_z = 0.8 gives us the root locus in Figure 5.5; this locus has a pair of poles with a damping factor slightly greater than 1, and pole values close to 0.7 (marked on the plot with a '+').

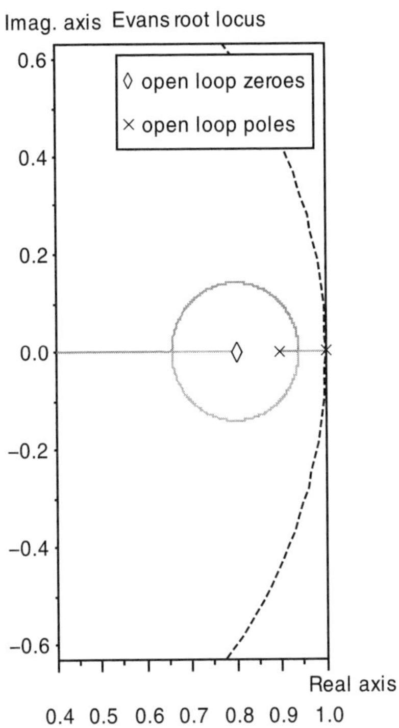

Figure 5.5 The updated root locus

The Evans root locus plot has a wider application than systems with a single variable gain. Take (5.10) or (5.11) from Example 5.1. In each equation there are *two* parameters available for the control engineer to vary. For both of these equations you can hold one parameter constant and vary the other to see the root locus.

There was an innocent-looking line in the directions for constructing a root locus plot that bears some explaining. Identifying the correct transfer function to do an Evans root locus plot *on* is usually straightforward, but sometimes it isn't trivial, or at least it isn't as trivial as it seems. The most reliable (but often most tedious) way of identifying the correct transfer function is to calculate your system's transfer function and group the denominator into a sum of two polynomials: one that's by itself, and the other that's multiplied by the parameter you want to vary. This is the method I used in Example 5.2. If your root locus is going to be made by varying a gain in the outer loop, then you identify the open-loop transfer function and use that—but it can be tricky in some cases to do this.

Take, for instance, the control loop shown in Figure 5.6. Such a system may be used where you have a velocity-controlled inner loop wrapped by a position loop. The closed-loop gain for this system is

$$G(z) = \frac{(k_1 k_2) z}{z^3 + (k_1 k_3 - 2) z^2 + (1 + k_1 k_2 - 2k_1 k_3) z + k_1 k_3}. \tag{5.12}$$

If you need to generate a root locus plot for the case where you are varying k_1 then the open-loop gain is just the gain of the block diagram shown in Figure 5.7, and the open-loop gain to use is

$$G_{ol}(z) = \frac{(k_1 k_2) z}{z^3 + (k_1 k_3 - 2) z^2 (1 - 2k_1 k_3) z + k_1 k_3}. \tag{5.13}$$

What if you need to find out what happens as you vary k_2 or k_3? You cannot use the forward path shown in Figure 5.7, because both of the constants are embedded in the loop. You can still create an Evans root locus as long as your transfer function is linear in the parameter of interest (and if you can't create an Evans root locus, you can always create a general root locus). Take the case

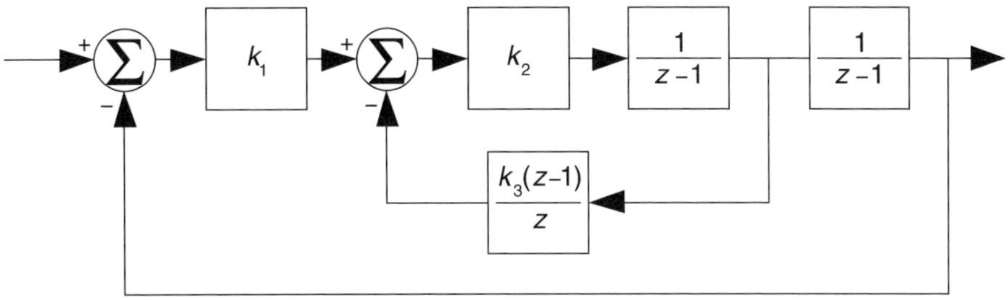

Figure 5.6 A two-loop control system

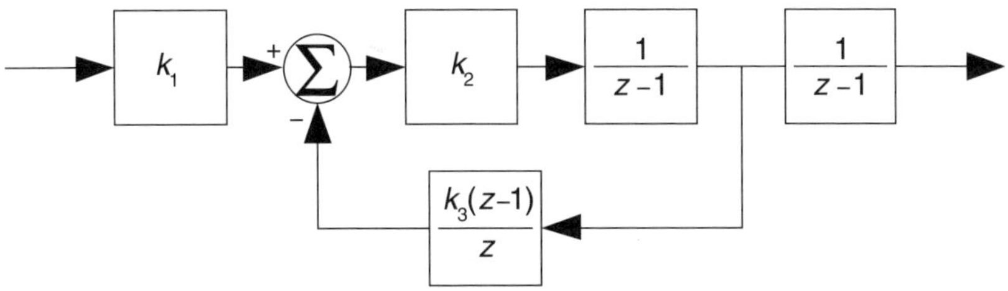

Figure 5.7 Block diagram to calculate open-loop gain of Figure 5.6

of k_2. To get the Evans root locus as k_2 is varied, you need to go back to (5.12) and rewrite it as

$$G(z)=\frac{(k_1k_2)z}{\left[z^3+(k_1k_3-2)z^2+(1-2k_1k_3)z+k_1k_3\right]+k_2k_1z}. \tag{5.14}$$

Notice that the denominator is separated into a part that is not dependent on k_2 and a part that is. You can use this as if you were dealing with a system where your open-loop gain was

$$G_{fol}=\frac{(k_1k_2)z}{z^3+(k_1k_3-2)z^2+(1-2k_1k_3)z+k_1k_3} \tag{5.15}$$

(where the 'fol' subscript denotes 'fictitious open loop'). Once you have arrived at this fictitious open-loop gain you can go ahead and plot the Evans root locus as normal.

Example 5.2 An Alternate Root Locus.

For the system in Example 5.1 find the root locus as d_z is varied with the gain held at $k_pA = 0.3$.

The overall transfer function of the system is

$$G(z) = \frac{G_{ol}(z)}{1+G_{ol}(z)} = \frac{k_pA(z-d_z)}{k_pA(z-d_z)+(z-0.9)(z-1)}. \tag{5.16}$$

Setting $k_pA = 0.3$ and simplifying we get

$$G(z) = \frac{0.3(z-d_z)}{z^2-1.6z+0.9-0.3d_z}. \tag{5.17}$$

or

$$G(z) = \frac{0.3(z-d_z)}{(z-0.8-j0.51)(z-0.8+j0.51)-0.3d_z}. \tag{5.18}$$

One can either deal with this equation by simply doing a root locus plot starting at $z = 0.8 \pm j0.51$, or one can make a fictitious open-loop transfer function that would generate the correct root locus if it were placed in a loop:

$$G_{fic}(z) = \frac{-0.3d_z}{(z-0.8-j0.51)(z-0.8+j0.51)}. \tag{5.19}$$

The resulting root locus plot is shown in Figure 5.8.

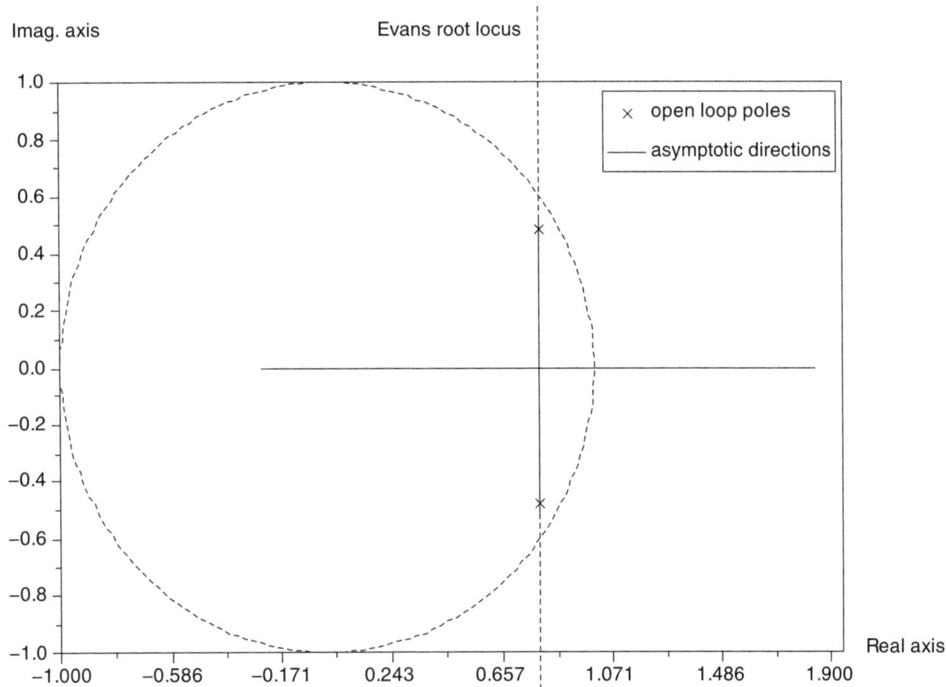

Figure 5.8 The Evans root locus plot

5.2 Bode Plots

The root locus suffers from three deficiencies: it takes a bit of experience to get a good feel for how close the system is to going "over the edge" into instability, it cannot deal with systems whose transfer functions cannot be expressed as rational polynomial ratios, and it does not provide a direct method for going from measured data about a real-life system directly to a control system design.

The Bode plot, introduced in Chapter 2, almost completely eliminates these difficulties at the expense of not giving you a road map of where your system is going as a parameter is varied. Instead, it shows you a detailed picture of where your system is for any given set of parameters, but it does so in a way that lets you estimate how "safe" the system is.

Determining Stability

In the case of the root locus plot, we were able to make a picture of the tracks of all of the system poles as some parameter was varied, essentially getting an aerial view of the possible pole locations as the parameters change. In the case of the Bode plot, we get a notion of how close the system poles are to being unstable, much in the same manner as a mariner would get an idea of water depth by taking soundings.

In both the Bode plot case and the root locus case, the critical value that we are looking for is the system changes necessary to bring the open-loop gain equal to −1. While this information is somewhat less direct with the Bode plot, it can be more immediately useful.

Assume that you've arrived at a closed loop system that is known to be stable, and assume that you've either calculated or measured the open-loop gain and phase response. You know that since the system is stable, the poles of the system must lie within the unit circle. Your job is to design a control system that keeps those poles inside the unit circle while achieving your other design goals.

For any system whose transfer function is parameterized and stable, if the coefficients of the system polynomial vary in a continuous manner, then the system's poles will change in a continuous manner.[2] This means that any parameter variation that will move a system pole *to* the unit circle will move that pole *out* of the unit circle with just a bit more variation.

A Bode plot is just a plot of the system open-loop gain as the value of z is scanned around the unit circle. Thus any change in a stable system that will bring the open-loop gain to exactly 1 will place one or more system poles on the unit circle, making the system unstable. This means that we can look at a Bode plot of a system's open-loop gain to see how close the system is to being unstable.

---

[2] This does *not* apply to all system polynomials. If you have a transfer function whose denominator is $az - 1$ then the system pole will jump from $+\infty$ to $-\infty$ as a goes from slightly above zero, through zero to slightly below zero—but the system is not stable while that is happening.

Traditionally the way that you express the robustness of a control system is by talking about its *gain margin* and *phase margin*. Both of these margins are measures of how close the open loop gain of the system is approaching –1. The goal of using these margins is to determine the amount of plant variation that can happen before the system goes unstable.

When a system has an open loop gain of 1 its gain is 0dB and its phase shift is 180°. Thus, at a frequency where the system's open loop phase shift is exactly 180° the system is only a gain change away from having a gain of exactly 1; the amount of gain change (in dB) required to get you to 0dB of gain at such a frequency is the gain margin. Similarly, at a frequency where the system's open loop gain magnitude is exactly one, the system is only a phase shift away from having a gain of exactly 1; the amount of phase change required to get you to 180° at such a frequency is the phase margin.

It would be nice if there were some absolute rule about how to set your gain and phase margins for ideal performance and robustness. System gain and phase margins *do* have a strong impact on the system's closed-loop frequency response and time-domain response, but you cannot say what margins can be considered safe without knowing how the plant (or controller) may vary.

Even while there are no absolute rules there *are* guidelines. In general, it's safe to say that at the lower end of your frequency response you can have gain and phase margins of 6dB and 60° and feel safe: a gain margin of 3dB or a phase margin of 30° would be pushing it.

As frequencies go up, however, you can expect a system to have a great deal more variation in its behavior. At frequencies that are much higher than your intended closed-loop bandwidth it is usually necessary to give more margin. Phase, in particular, can vary widely at the upper end of a system's operating frequency range, and it is often meaningless to speak of phase margin much above the loop closing frequency.

Example 5.3 Using the Bode Plot to Determine Gain and Phase Margins

The high-performance control system pictured in Figure 5.9

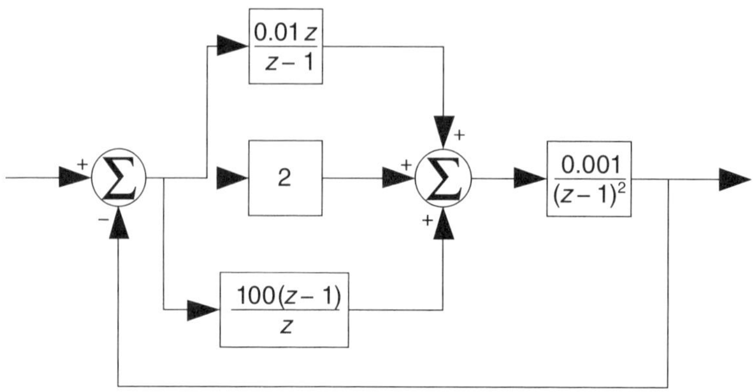

Figure 5.9 A control system

has a plant whose transfer function is

$$G(z) = \frac{0.001}{(z-1)^2}, \tag{5.20}$$

and a PID controller with transfer function

$$H(z) = 100\frac{z-1}{z} + 2 + 0.01\frac{z}{z-1}. \tag{5.21}$$

Find the open-loop frequency response and make a Bode plot, find the system's gain and phase margins, and show a Bode plot of the system closed-loop frequency response.

The system open-loop frequency response can be found by entering (5.20) and (5.21) into an appropriate math package and plotting the result, as shown in Figure 5.10.

The open-loop gain crosses the 0dB line at about 17Hz, with a phase of 110° for a phase margin of 70°. It crosses 180° at 1.6Hz and 160Hz, with gains at 26dB and –20dB, respectively, for a minimum gain margin of 20dB.

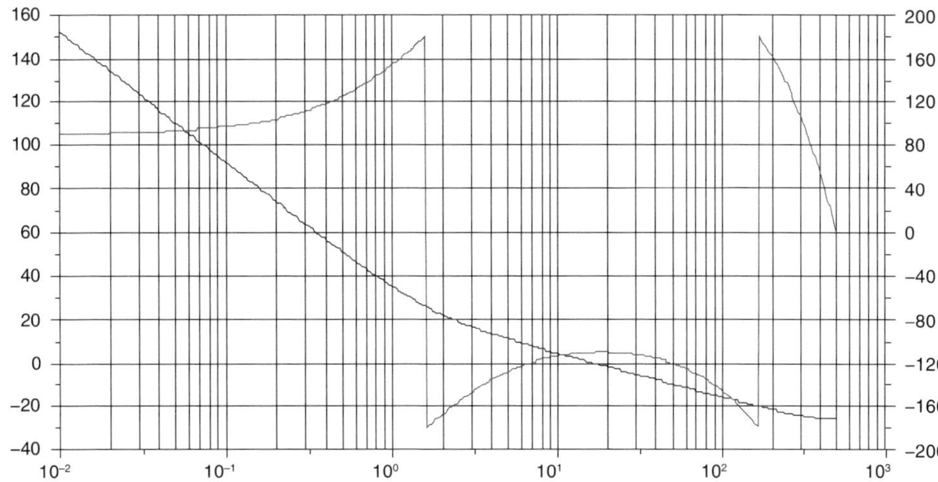

Figure 5.10 Open-loop bode plot of the system

Figure 5.11 is the system's closed-loop Bode plot. In it, you can see some minor peaking of the closed-loop response around ½Hz, but the frequency response is otherwise nice and flat, with a 3dB bandwidth of about 20Hz.

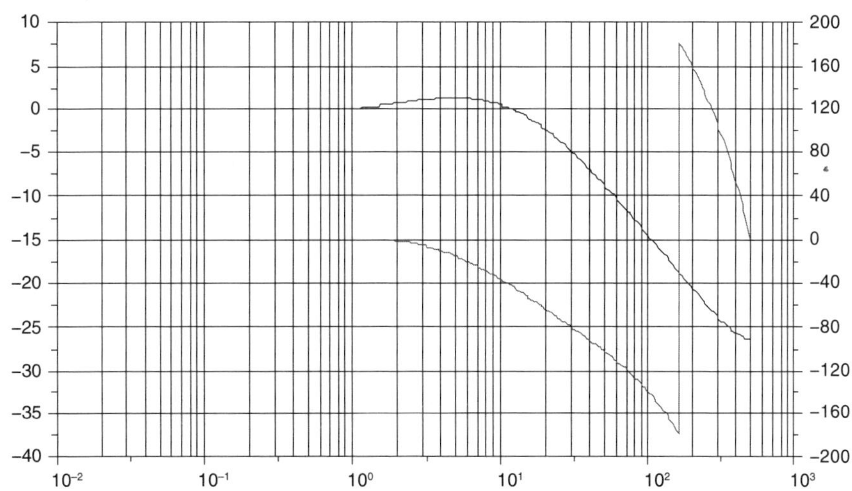

Figure 5.11 Closed-loop bode plot of the system

Sensitivity

All systems are subject to disturbances. Figure 5.12 shows such a system whose plant is being driven by both the controller output and a disturbance input.

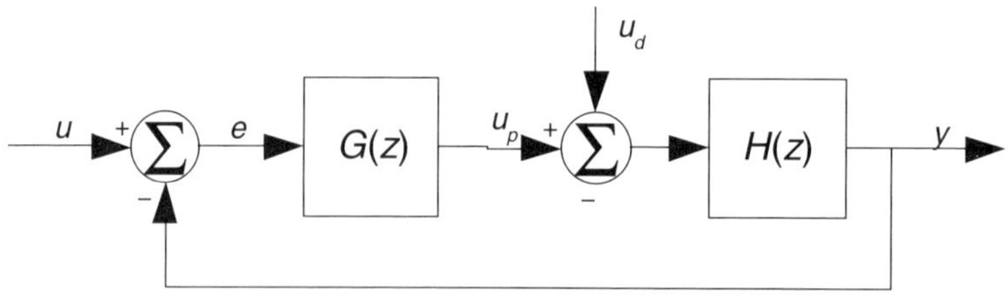

Figure 5.12 A system with a disturbance input

With a system such as the one in Figure 5.12, the system's response to a disturbance is

$$H_d(z) = \frac{H(z)}{1+G(z)H(z)} \tag{5.22}$$

This disturbance response can be thought of as the plant's own response multiplied by a *sensitivity*, or

$$H_d(z) = S(z)H(z), \tag{5.23}$$

where

$$S(z) = \frac{1}{1+G(z)H(z)} \tag{5.24}$$

is the sensitivity, which is completely dependent on the system's open-loop gain.

Anywhere the sensitivity of the system is lower than one, the system performance will be improved over just using the plant alone. Conversely, anywhere the sensitivity of the system is *greater* than one, the system performance will be degraded. It has been shown[3] that you cannot reduce the sensitivity of the

[3] Bode's sensitivity integral

system at one frequency without increasing it at another. In the case of control systems, this means that any improved sensitivity that we achieve must be paid for by degraded sensitivity somewhere else.

A system's gain and phase margin have a direct connection to its sensitivity. At any frequency that the open-loop phase shift is 180° and the open-loop gain magnitude is less than one the system sensitivity will be greater than one; similarly at any frequency where the open-loop gain magnitude is 1 and the phase margin is less than 60° the sensitivity will be greater than one. So a system with a low gain or phase margin will not only be close to being unstable, it will also have "peaks" to its frequency response.

Because a system's sensitivity can peak around low gain or phase margins, it is a good idea to pay attention to your margins when you are designing. Because you cannot reduce sensitivity at a frequency without increasing it somewhere else, you have to pay attention to where the extra sensitivity is going—almost always when you are designing a system, you care about the system's behavior at lower frequencies, and less at high ones. Thus, it is almost universal practice for control systems to pay for better performance at DC and low frequencies by sacrificing it at higher frequencies. Any time your system must reject disturbances, you should pay careful attention to the extra sensitivity you are generating.

5.3 Nyquist Plots

A Bode plot shows a system's phase and gain as separate quantities, plotted along the frequency axis. A Nyquist plot uses the same data as a Bode plot, but it discards the frequency axis to the track traced by the system's frequency response on the complex plane. This doesn't allow you to see the frequency data without annotating the plot, but it *does* allow you to see at a glance the gain and phase margins of the system, as well as the system's peak increase in sensitivity.

Figure 5.13 shows a Nyquist plot of the open-loop system response shown in the Bode plot in Figure 5.10. As with any other frequency design method, the important consideration for both robustness and stability is how closely the system's complex gain approaches 1.

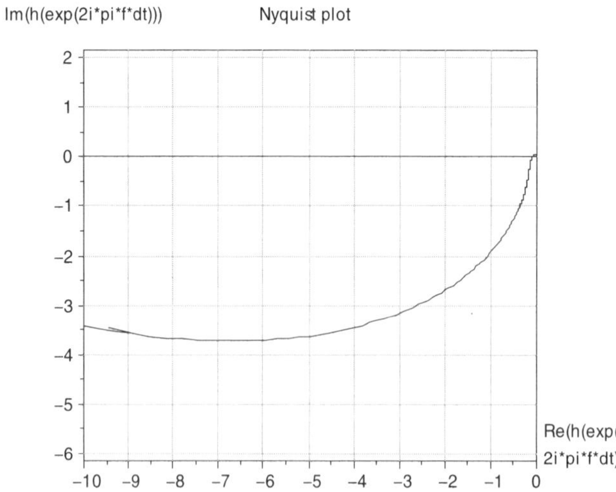

Figure 5.13 A Nyquist plot of the response plotted in Figure 5.10

Determining Stability

One can take two approaches to determining a system's stability with the Nyquist plot. Like the Bode plot, a Nyquist plot will tell you how closely a system approaches instability and where its sensitivity is poor. It will, in fact, do a better job than the Bode plot by giving you a direct reading of the maximum sensitivity without having to try to interpolate between specific gain and phase crossings. Unlike the Bode plot, the Nyquist plot allows you to directly determine a system's stability just by looking at the shape of the plot.

In the first case, you take the Nyquist plot to be a variety of Bode plot where the frequency isn't shown and the phase and gain are determined simultaneously. This is a very useful way of approaching the Nyquist plot, but aside from the direct reading of system sensitivity it just reiterates the Bode plot's capability of determining when a previously stable system will go unstable.

With a bit more work, you can use the Nyquist plot, plus some prior knowledge about the system structure, to determine the system stability "directly" without starting from a known stable system.

With the Bode plot style of assessing system stability, you can see how closely the system is to going unstable, but only if you are starting with a known-stable system. With the "direct" method of determining system stability you do not have to do this. In practice, however, this "direct" method is more indirect to apply. It can be difficult to keep all of your 'i's dotted and your 't's crossed while using it, and it requires information that you won't always have available. Nonetheless it is sometimes useful, and gives an interesting view of system stability.

The method uses a mathematical property of polynomials called "the principle of the argument," which I will not prove here. The principle of the argument states that if you have a polynomial $P(z)$ and you trace z along any closed path that doesn't touch the roots of $P(z)$, then the value of $P(z)$ will form a trace on the complex plane that circles 0 exactly once for each of the roots of $P(z)$ that are enclosed in the path. Furthermore, if the most significant coefficient of $P(z)$ is positive and the path traced by z is in a counter-clockwise direction, the trace of the values of $P(z)$ will also be counter-clockwise.

Figure 5.14 illustrates how the principle of the argument works. A random third-order $P(z)$ was selected, as well as an arbitrary path for z to follow. The left side of Figure 5.14 shows the root locations for $P(z)$ and the path followed by z; the right side of Figure 5.14 shows the trace of the values of $P(z)$ as z is varied along the path shown on the left. The trace of values encircles the origin exactly three times, once for each root of $P(z)$.

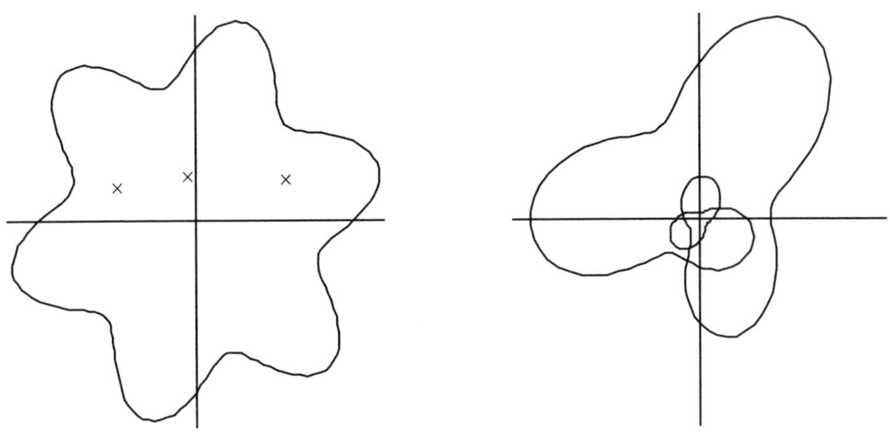

Figure 5.14 Using the principle of the argument

It is easy to see that if the values of $P(z)$ follow a trace that encircles the origin a certain number of times, and if $P(z)$ never goes to zero, then $1/P(z)$ must also encircle the origin the same number of times, but in the opposite direction. Furthermore (and again without proof), if $H(z)$ is a ratio of polynomials in z, then as z is varied along its path the trace of $H(z)$ will circle the origin exactly $Z-P$ times, where Z is the number of zeros enclosed by the path and P is the number of poles enclosed by the path.

As pointed out before, when you find the frequency response of the system, you are tracing z along the upper half of the unit circle and capturing the resulting values of $H(z)$. Because $H(z)$ describes a real system, all of its poles and zeros are either real-valued or come in complex conjugate pairs. As a consequence, the trace of $H(z)$ is symmetrical around the horizontal axis.[4]

If you know the total number of poles in the system and the number of stable poles then you can compare the two and easily see if there are any unstable poles. Inspecting the trace of $H(z)$ will tell you how many stable poles there are compared to the number of stable zeros. So there are three pieces of information you would need to use the principle of the argument to find if a system is stable: the total number of poles, the number of stable zeros, and the trace of $H(z)$. Assuming that you have this information in hand, you can use the "direct" method to find the system stability.

Example 5.4

A system has the transfer function

$$H(z)=\frac{(z-0.8)(z-1.2)}{z^3-2.5z^2+2.12z-0.612}. \tag{5.25}$$

Use the principle of the argument and the frequency response of $H(z)$ to find if the system is stable.

---

[4] You can, of course, simply trace z around the entire unit circle—but doing so can obfuscate which part of the resulting Nyquist plot comes from which part of the frequency response.

Figure 5.15 shows the trace of $H(z)$; the heavy portion of the trace was calculated as z ranged from 0 to π (the top half of the unit circle), while the light portion of the trace is simply a reflection of the solid portion. The start of the trace is denoted by an 'x' to allow you to see direction.

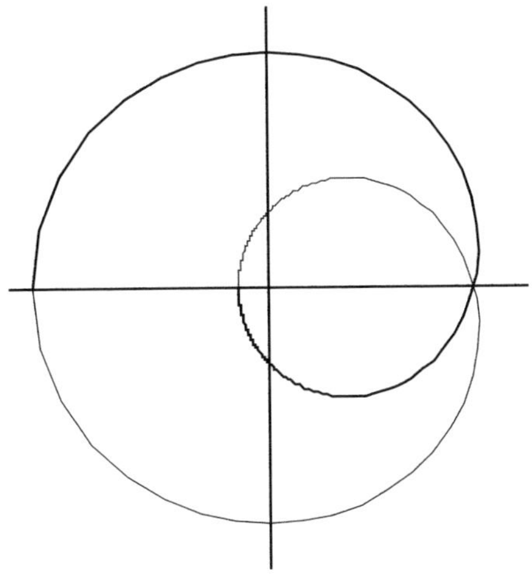

Figure 5.15 The trace of H(z)

Clearly $H(z)$ has one stable zero (at $z = 0.8$), one unstable zero (at $z = 1.2$) and three poles. Because the frequency response of $H(z)$ gives the result as if z were traced counterclockwise on the unit circle $H(z)$ would be stable if the full frequency response of $H(z)$ rotated around the origin twice, in a clockwise direction.

If you examine Figure 5.15, you will see that the trace does indeed circle the origin twice in the clockwise direction. Therefore you can conclude that the system is indeed stable.

But how does the principle of the argument relate to the Nyquist plot? To see how, consider the system shown in Figure 5.16.

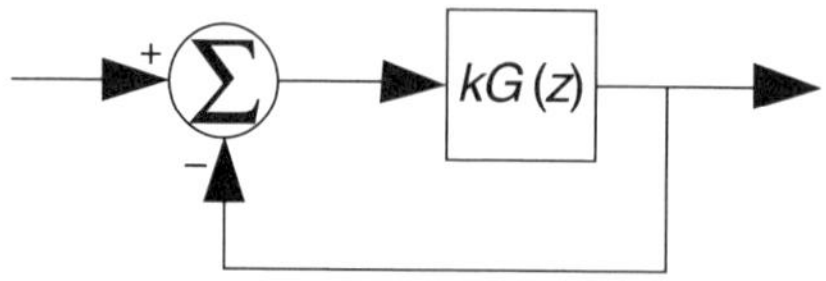

Figure 5.16 A control system

The closed-loop transfer function for this system is

$$H(z) = \frac{kG(z)}{1 + kG(z)}. \tag{5.26}$$

The stability of $H(z)$ is determined solely by the open-loop gain $kG(z)$. Specifically, if the transfer function

$$S(z) = \frac{1}{1 + kG(z)} \tag{5.27}$$

is stable then $H(z)$ must also be stable. The transfer function $S(z)$ is stable only if all the zeros of

$$1 + kG(z) \tag{5.28}$$

are stable,[5] because the zeros of (5.28) are the poles of both S and H. To find the number of stable zeros in (5.28) we can use the principle of the argument, in this case simplified by the fact that in all but one very peculiar case, the number of zeros in (5.28) is equal to the number of poles. So, finding the number of *stable* zeros boils down to knowing the number of unstable poles in $G(z)$, and using the principle of the argument to verify that the trace of (5.28) circles the origin in a counterclockwise direction once for each unstable pole of $G(z)$.

---

5 The term "unstable zero" means a zero that lies outside of the unit circle. The term can be misleading because a system can have any number of unstable zeros and still be perfectly stable—it is industry standard terminology, however, and does make sense once you get used to it.

The Nyquist plot can be used for this task with two adjustments to our procedure. The first adjustment is simply to observe that the Nyquist plot deals with $kG(z)$, but (5.28) is $1 + kG(z)$. This is is a minor difficulty that is easily dealt with: if (5.28) circles the origin, the Nyquist plot of $kG(z)$ circles -1.

The second adjustment to our procedure is to observe that many controllers and some plants are integrating, with poles at $z = 1$. This puts the poles exactly on the boundary where we are scanning z, which both violates the assumptions of the principle of the argument and causes the frequency response of $G(z)$ to go to infinity at DC. Also, you will occasionally have a plant with resonant pole pairs exactly on the unit circle, which causes similar issues at the resonant frequency.

When one is generating Nyquist plots from mathematical models, one can sidestep the issue of poles on the unit circle by changing the path followed by z, diverting it around the trouble spot, then back to the unit circle.

Example 5.5 Using the Nyquist Plot to Determine Stability

A system has an open-loop transfer function of

$$G(z) = \frac{(z-0.76)(z-0.64)}{(z-1)(z^2-1.4z+0.58)}. \tag{5.29}$$

Use the Nyquist plot to find if this system is stable, avoiding the pole at $z = 1$.

Figure 5.17 shows a possible track for z for the system.

Figure 5.18 shows the Nyquist plot resulting from using the path shown in Figure 5.17. The large semicircular section is the portion of the plot resulting from sidestepping the pole at $z = 1$, the relatively straight portions show where the plot would asymptotically go to infinity, the straight dotted lines show the plot going off to infinity, and the balance of the plot is just the "normal" Nyquist plot for this system.

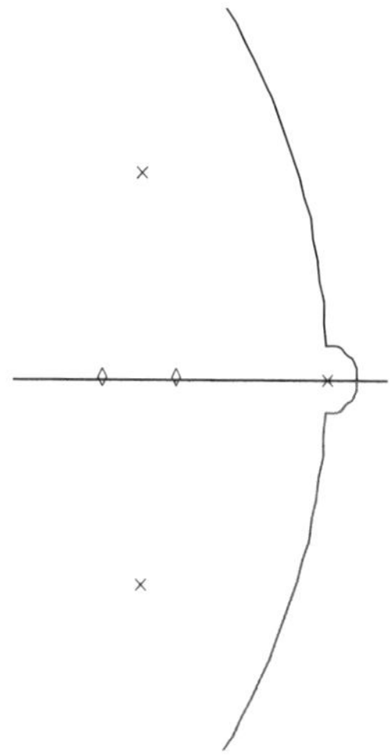

Figure 5.17 Path for z to avoid a pole at z = 1

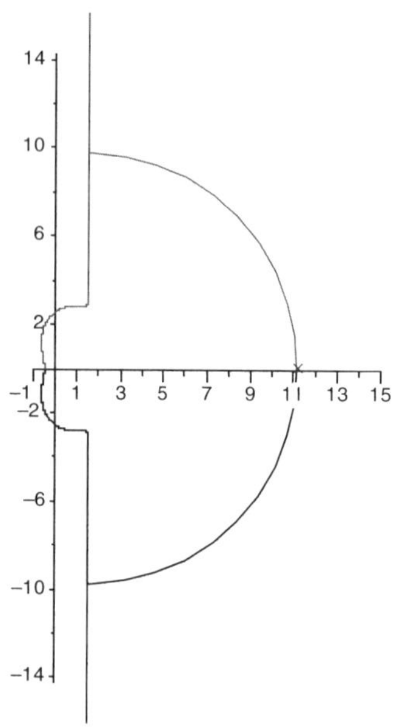

Figure 5.18 Nyquist plot resulting from z taking the path in Figure 5.17

Because the track taken by the Nyquist plot does not circle $G(z) = -1$ at all, the system is stable.

When one is using measured data to generate Nyquist plots, one does not have the luxury of choosing to sidestep poles on the unit circle as was done in Figure 5.17 and Figure 5.18. In addition, if the system is unstable, it is not always clear from measurements alone how many unstable poles the system possesses.

The first problem can be cleared up by looking at the direction of departure of the plot as the gain goes to infinity and judge in which direction and how many times the plot would circle if the plot could have been generated theoretically. This is illustrated in Figure 5.18, where the frequency response is continued out to infinity as the pole at $z = -1$ is approached. It is still necessary to know if the Nyquist plot at DC represents one pole or five; an experienced reader of Nyquist plots can tell this by looking at how the phase evolves, but one can always determine this number fairly quickly by looking at the amplitude response of the Bode plot—as the frequency goes toward DC, the gain will rise at 20dB of gain for every decade of frequency for each pole—a 20dB/decade rise would indicate one pole, 40dB/decade would indicate two, etc.

The second problem can be dealt with in one of two ways: either do some theoretical modeling of the system to determine the number of unstable poles, or stabilize the system, however poorly, then measure its frequency response in that state and use the number of times the –1 spot is circled by the open-loop response to determine the number of open-loop unstable poles.

Of course, once you've managed to stabilize the system, you can also sidestep both of the above problems. In this case, the task of designing a control system to a Nyquist plot simply becomes one of establishing a "keepout" zone around $G(z) = -1$ and designing one's controller accordingly.

Sensitivity

Look back to the system shown in Figure 5.12, where we have defined the sensitivity of the system in the context of the Bode plot. I pointed out in that section that one could only interpolate sensitivities between gain crossing and phase crossing frequencies.

The Nyquist plot is different. A very handy feature of Nyquist plot is that it is easy to get a feel for the minimum magnitude of the loop response. When you generate a Nyquist plot, you are generating a plot of the open-loop gain, which is directly related to the sensitivity. In fact, you can restate (5.24) as

$$S(z)=\frac{1}{1+G_{ol}(z)} \tag{5.30}$$

where G_{ol} is the open-loop gain. This means that the sensitivity of the system at any point is just the reciprocal of the distance between 1 and G_{ol} at the frequency in question. Thus, you can easily see the maximum sensitivity of the system by looking for the closest approach of the trace to 1. In Figure 5.13 a circle of radius ½ has been inscribed on the plot, indicating a maximum sensitivity of two times (6dB) the "natural" plant sensitivity. Such a circle, at whatever diameter you deem appropriate, is a useful aid to graphical system design, where you can adjust the controller parameters while keeping an eye on whether the open-loop gain is violating some predefined sensitivity limit.

Gain and Phase Margins

Determining system gain and phase margins is fairly simple with the Nyquist plot. Recall that the gain margin is the amount of gain change needed to bring the loop gain to 1 when it is a pure negative number, and that the phase margin is the amount of phase change needed to bring the loop gain to 1 when the absolute value of the loop gain is 1.

The gain margin of a system is the gain at which the phase of the open-loop response is 180 degrees. On the Nyquist plot this occurs any time the response crosses the negative real number line. Figure 5.19 shows a pair of Nyquist plots made from the system of Figure 5.10 which show the system gain margins. These plots are identical to Figure 5.13, except that the scaling has been changed in each case to show the zero crossing clearly. The plot on the left is a high-gain plot, showing the loop gain crossing the real number line at a gain of about 20 (26dB), while the plot on the right shows the loop gain crossing the real number line at a gain of about 0.1 (–20dB).

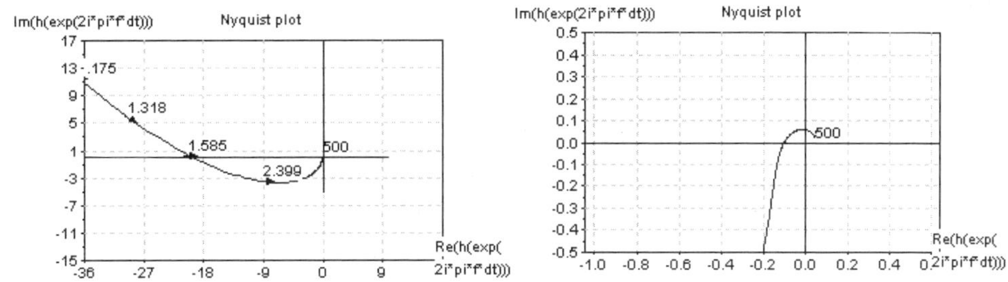

Figure 5.19 Nyquist plots scaled for gain margin

The phase margin of a system is the difference between the phase of the open-loop response and 180 degrees, when the open loop gain amplitude is equal to one. Figure 5.20 shows a Nyquist plot with a circle of radius 1 inscribed. The straight line indicates the phase angle of the loop gain where it crosses this “gain = 1” circle. The phase angle is 110 degrees, for a phase margin of 70 degrees.

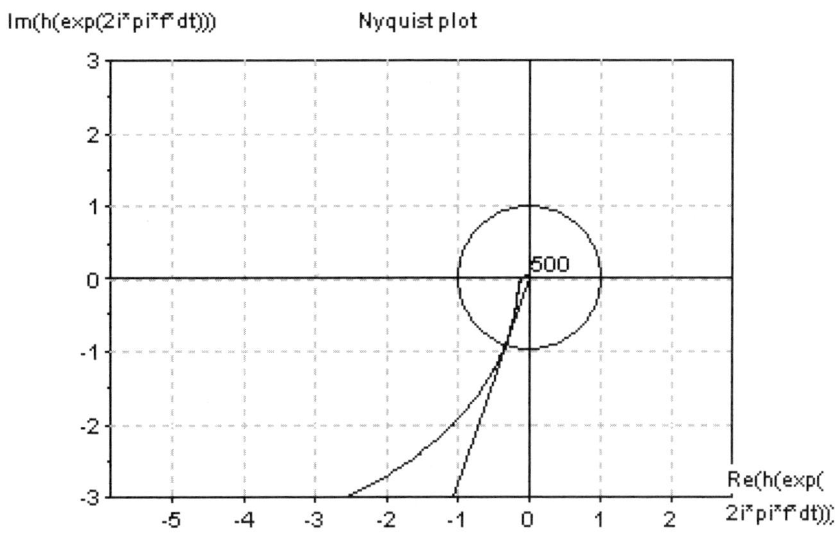

Figure 5.20 Nyquist plot showing phase margin

5.4 Conclusion

This chapter has shown three popular and useful methods for analyzing control system designs. Each one of these methods has its own particular advantages, and each one of these methods can be used to illuminate a particular aspect of a system's behavior and how its behavior will vary if its characteristics change.

Because none of these methods are complete (and in fact the root locus method will only work if you have a z-domain model of the system), one should make a habit of using a combination of these methods in one's controller design.

6

Design

Previous chapters have brought us to the point where, given a system design, we can look at it every which way, analyze and specify its performance, and communicate its structure and behavior.

This chapter shows how, once you have a plant chosen and a system specified, you can design a controller to attain your control system goals.

6.1 Controllers, Filters and Compensators

When a control engineer talks about a *compensator*, he is referring to an element of a control system, usually a filter, that corrects or otherwise deals with an element of the plant's behavior. Indeed, if the discussion is limited to linear control theory then the *only* kind of compensator that enters the discussion will be a linear filter. For instance, if a plant cannot, by itself, guarantee zero steady-state error then an *integral* or *reset* compensator may be introduced. If a plant has some nonlinear behavior that can be easily reversed with a cleverly placed inverse function in the controller, then such a block would be called a *nonlinear compensator*.

Nearly any block in a controller that has a single input and a single output can be called a filter—even a simple gain block is a filter in a very simplistic sense. The term filter has a reasonably exact meaning in linear systems theory, so a linear controller can be thought of as nothing more than a large, complex filter. A better use of the term "filter" in control systems, however, would be for elements of the system that precondition a signal before it is used, and that don't significantly affect the nature of the signal within the bandwidth of the control system.

When a set of filters and compensators are assembled into a system that controls the plant, the result is a *controller*.

6.2 Compensation Topologies

Cascade Compensation

Cascade compensation is perhaps the most common control system topology. With cascade compensation the error signal is found, and the control signal is developed entirely from the error signal. Figure 6.1 shows an example of cascade compensation where G is the plant and H is the controller.

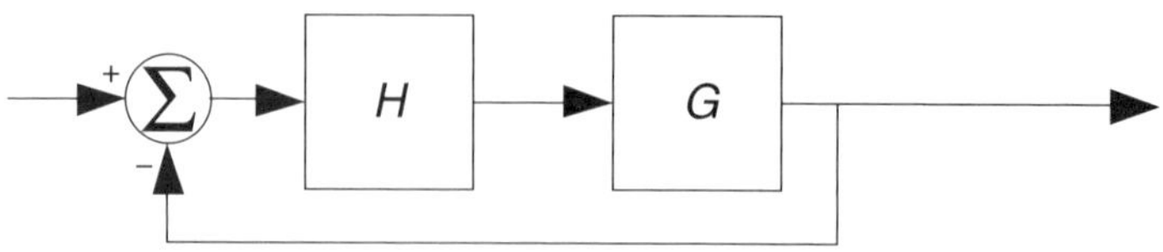

Figure 6.1 Cascade compensation

In cascade compensation, the system transfer function is

$$H_s = \frac{HG}{1+HG}, \tag{6.1}$$

with the salient characteristic that the entire controller characteristic appears in both the numerator and denominator of the transfer function—so anything you do to affect the system poles also affects the system zeros.

Feedforward Compensation

Feedforward compensation is shown in Figure 6.2. Feedforward compensation injects some of the command signal into the control signal without being affected by feedback. It can provide a way to speed up a system's transient response without affecting its pole locations or impacting its stability.

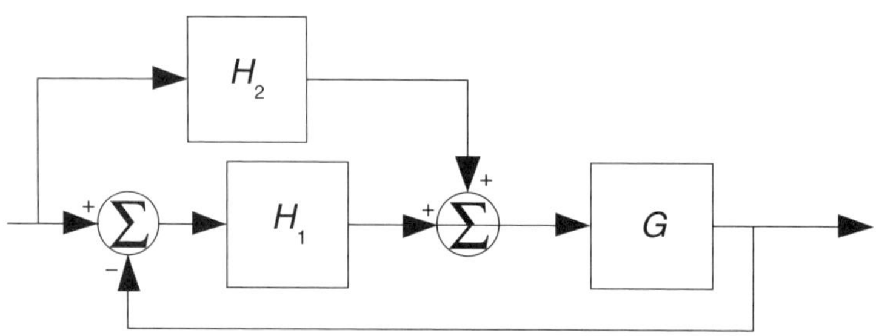

Figure 6.2 Feedforward compensation

The system transfer function for the feedforward compensation scheme shown in Figure 6.2 is

$$H_s = \frac{(H_1 + H_2)G}{1 + H_1 G}. \tag{6.2}$$

The balance between the amount of cascade compensation and feedforward depends on your goals and your plant. Given a good enough plant and motions that must happen as fast as possible you could set H_1 to zero and simply use open-loop control through H_2. On the other hand, given a poor plant or a regulation application where you care about rejecting disturbances but don't much care about following a command, you could set $H_2 = 0$ and revert entirely to cascade compensation. Often if you need to use a feedback controller at all you don't need or can't use much explicit feedforward, although it can be a useful tool to have.

Compensation in Feedback

Feedback compensation, shown in Figure 6.3, modifies the plant output in some way before it is compared to the command signal. This is generally done because the property that we wish to control is not the one that we are measuring; for example, if we want to control the velocity of a system with position measurement. Showing a functional block in the feedback path is also a good way to model how the measurement process affects the signal. Whether it is a simple gain block to model an analog-to-digital converter, or something more complex such as a gyroscope or an accelerometer, you can model the behavior of a sensor with a feedback block.

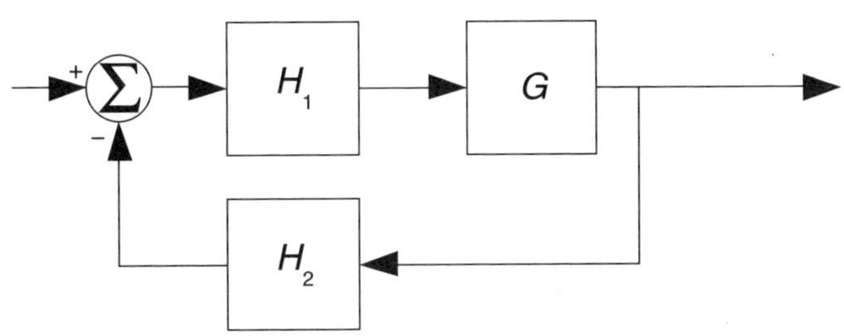

Figure 6.3 Feedback compensation

The transfer function of a system with feedback control is

$$H_s = \frac{H_1 G}{1 + H_1 H_2 G}. \tag{6.3}$$

Putting compensation in the feedback path of a system, either explicitly in the form of a filter, or implicitly in the form of a sensor, can have a dramatic effect on the "personality" of the system. Instead of the system attempting to servo to the system output, it will now attempt to servo to the output of the feedback compensation; thus, a rotary system with a gyroscope or tachometer in the feedback path will tend to servo the plant's rate of rotation, while a system with an accelerometer will tend to servo to the plant's acceleration. Careful use of feedback compensation, particularly by sensor selection, can be very powerful indeed.

6.3 Types of Compensators

This section will talk about the various types of compensators that are popular in industry, discuss how they might be implemented and show how they are used to increase system performance.

Proportional

A proportional controller doesn't really deserve the term "compensation." Proportional-only loops can be quite useful, however, and they are certainly easy to implement and describe. By itself the proportional controller is often of limited usefulness, but one can occasionally use a purely proportional controller and when a single control element can be used, it is most likely to be a proportional element.

Example 6.1

An aircraft's elevator is trimmed by driving the angle of its horizontal stabilizer with a DC motor driving a jackscrew through a gearbox. The motor/gear assembly has a mechanical time constant of 100ms, and the control system is sampled at 100Hz. The controller must be able to settle in no less than 1 second to within 5% of its final value. Use an approximate plant model to design a controller, and verify that proportional compensation will be adequate for the task.

The motor will act like a low-pass filter followed by an integrator. An approximate motor model can be formulated, knowing the mechanical time constant:

$$\frac{X}{U} = \frac{0.001 k_m z}{(z-1)(z-0.9)}, \tag{6.4}$$

where u is scaled such that $u = 1$ is full-scale drive, and k_m is the jackscrew speed at full drive.

Figure 6.4 shows the Evans root locus plot of the system as the gain is varied from 0 to $5000/k_m$. Figure 6.5 shows the same plot zoomed in to the vicinity of $z = 1$, where we can see that the system root loci intersect at about $z = 0.95$. This corresponds to a system time constant of 200ms, or a settling time of about 600ms, so we know that a pure proportional controller will be comfortably within specifications for settling time.

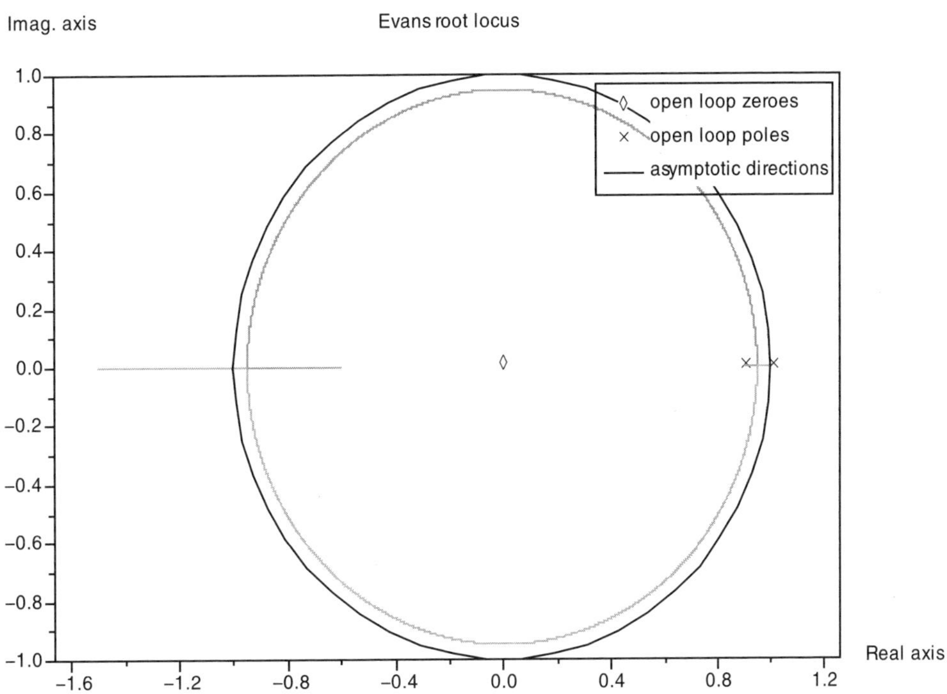

Figure 6.4 Root locus plot of system

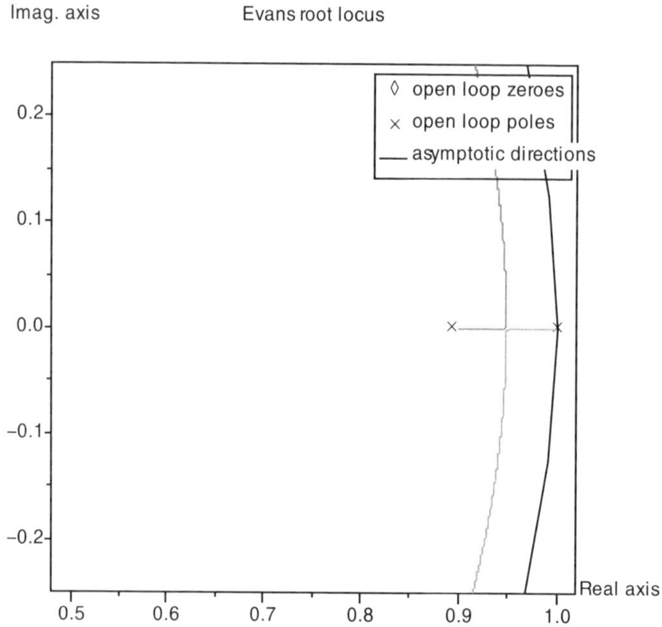

Figure 6.5 Root locus plot of system, zoomed in

Integral

Like proportional control, integral compensation is rarely used alone. Integrators have the nice property that as long as the average error at their input is nonzero, they will keep incrementing their output until the average error *does* reach zero. In frequency domain terms, this translates to the integrator having infinite DC gain, which forces the DC error to zero.

There is a price to be paid for this advantageous behavior at DC: the integrator output lags its input by a constant 90 degrees in the frequency domain. This lag reduces the phase margin of the system, and can easily lead to an unstable system.

Generally, when integral control is used, it is applied in parallel with proportional control, such as

$$H_c(z) = k_p + \frac{k_i}{z-1}, \tag{6.5}$$

which is a classic description of a proportional-integral (PI) controller. Figure 6.6 shows a Bode plot of the PI controller in (6.5) with $k_p = 10$ and $k_i = 0.1$. This controller has the advantage of a high gain at DC, which we want, while having a phase shift at higher frequencies where we may be closing the loop.

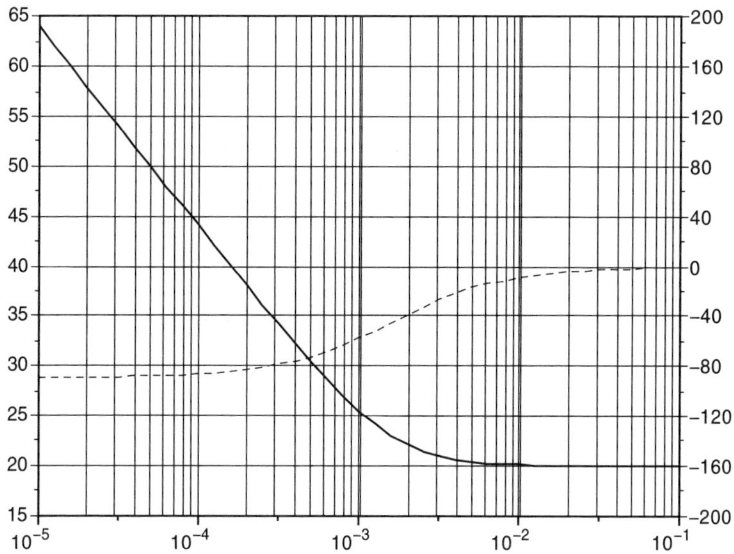

Figure 6.6 Bode plot of a PI controller

As an example consider a medical system where a blood pressure cuff is to be filled with a DC motor driving a pump. The pump driving into the backpressure of the tank forms a low-pass filter with a very low cutoff frequency. Additionally, the pump and motor have friction, which can be modeled as an unpredictable offset to the input voltage. The system samples at 20Hz. The plant model, with voltages and pressures normalized to the range [0, 1) is shown in Figure 6.7, where u is the control input, and u_f is the unknown offset.

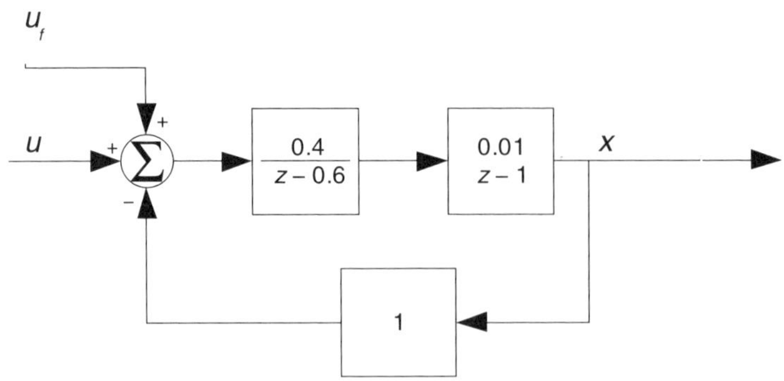

Figure 6.7 The pressure cuff plant model

A common task for a plant like this would be to design a controller that brings the pressure cuff to the correct value regardless of the offset voltage.

The transfer function from the intended input to the output for this system is

$$\frac{X}{U} = \frac{0.004}{z^2 - 1.6z + 0.604}. \tag{6.6}$$

This transfer function has a DC gain of 1. The transfer function from the voltage offset to the output is the same, with the same nonzero DC gain—it is this nonzero gain from an unpredictable error source that prevents the system from being realizable with a simple proportional controller.

The transfer function in (6.6) is very similar to the transfer function of the motor/gear system in Example 6.1. From this, we can conclude that we can add proportional gain to the system to gain speed. Figure 6.8 is a Bode plot of the pressure cuff system. On inspection, you can see two things: first, if just integral control were added with its 90 degree phase shift, the loop could not be closed at a frequency higher than about 0.02Hz, which would result in a very long settling time; second with proportional control a phase shift of –120° (60° phase margin) is reached at 0.6Hz, with a gain of –26dB (0.05). This indicates that a proportional controller with a gain around 20 should give a good response.

Figure 6.9 shows the form that the system takes with proportional feedback. The transfer function from the command signal to the output is now

$$\frac{X}{U_c} = \frac{0.004k_p}{z^2 - 1.6z + 0.604 + 0.004k_p}. \tag{6.7}$$

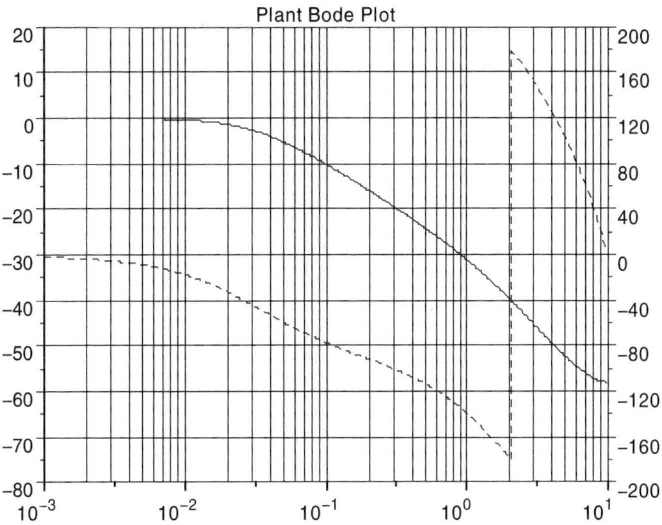

Figure 6.8 Bode plot of pressure cuff plant

With $k_p = 15$ this becomes

$$\frac{X}{U_c} = \frac{0.06}{z^2 - 1.6z + 0.664}, \tag{6.8}$$

which has poles at $z = 0.8 \pm j0.155$, and has a DC gain of 0.06/0.064. Note that with proportional feedback the DC gain of the system *will* be decreased when feedback is added. This is an unavoidable consequence of proportional feedback.

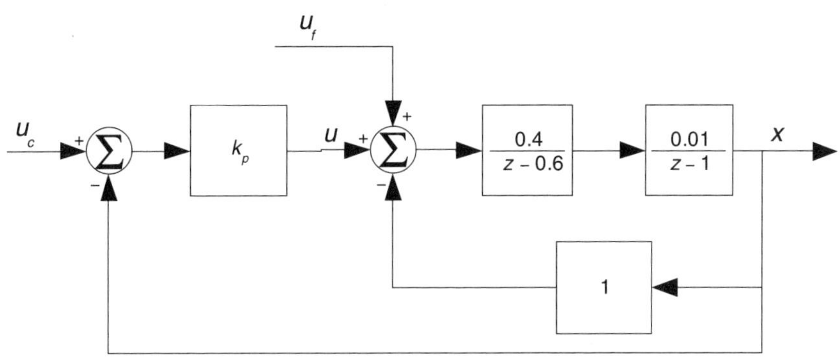

Figure 6.9 The pressure cuff with proportional control

Figure 6.10 shows the time domain response of the plant with a pure linear controller added, with gains of 20 and 15. The response with a gain of 20 overshoots by 10%; we'll choose a gain of 15 initially. Notice that neither of these responses reach the target value—this is due to the fact that after we add feedback the DC gain of the system with proportional control isn't quite 1.

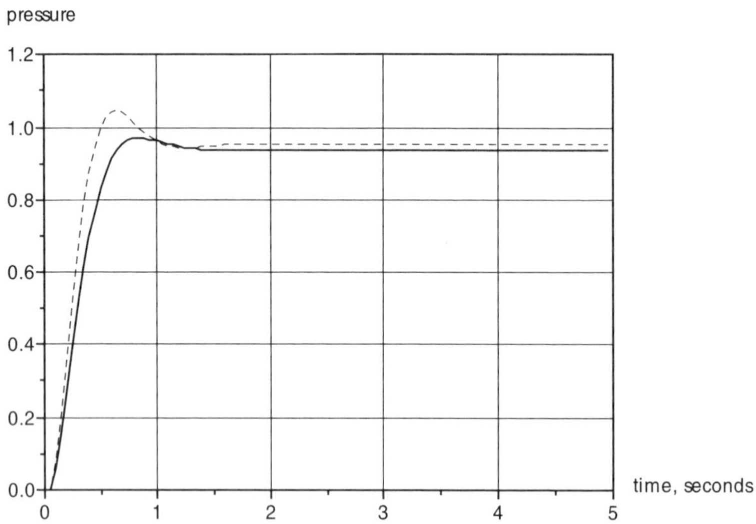

Figure 6.10 Time response of cuff with proportional control

If we take the results shown in Figure 6.8 and Figure 6.10 we can see that we should be able to add an integrator to the system with a good degree of safety. This can be double-checked with a root locus plot. First, find the transfer function of the system with k_p in Figure 6.9 replaced with a PI controller:

$$\frac{X}{U_c} = \frac{\left(k_p + \frac{k_i}{z-1}\right)\left(\frac{0.004}{z^2 - 1.6 + 0.604}\right)}{1 + \left(k_p + \frac{k_i}{z-1}\right)\left(\frac{0.004}{z^2 - 1.6 + 0.604}\right)} \tag{6.9}$$

which simplifies to

$$\frac{X}{U_c} = \frac{0.004\left((z-1)k_p + k_i\right)}{0.004\left((z-1)k_p + k_i\right) + (z-1)\left(z^2 - 1.6z + 0.604\right)}. \tag{6.10}$$

We can isolate the effect of k_i to get

$$\frac{X}{U_c} = \frac{0.004\left(\left(z-1\right)k_p + k_i\right)}{z^3 - 2.6z^2 + \left(0.004k_p + 2.204\right)z - 0.604 - 0.004k_p + 0.004k_i}. \tag{6.11}$$

The resulting root locus is shown in Figure 6.11; from this, we can see that the system should have a wide stability range as the integrator gain is varied.

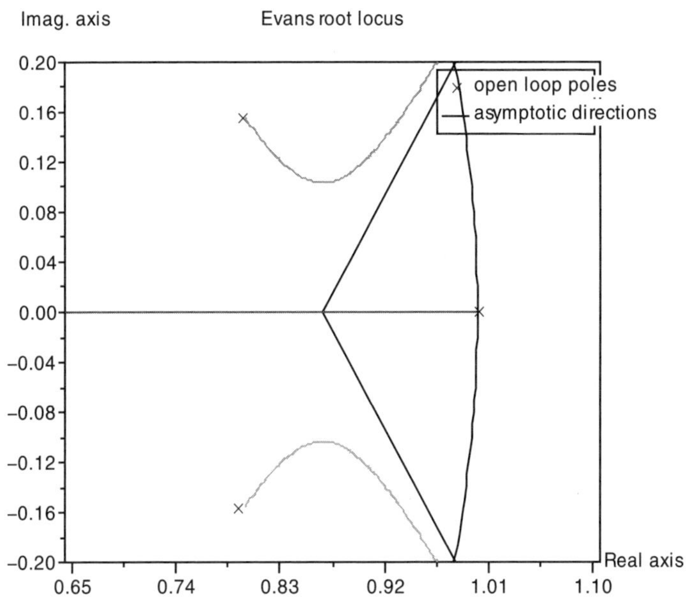

Figure 6.11 Pressure cuff integral root locus

We want to set the integrator gain low enough that it won't have a great deal of effect at the gain crossover frequency. If we choose an integral gain such that the integral action is 1/10th the value of the proportional action at the crossover frequency then the phase shift will be about 5 degrees:

$$\left|\frac{k_i}{z-1}\right|_{z=e^{j2\pi\frac{0.0.6Hz}{20Hz}}} = \frac{k_p}{10} \tag{6.12}$$

which works out to

$$k_i = \left|\frac{e^{j0.2}-1}{10}\right| k_p \simeq 0.02k_p = 0.3 \tag{6.13}$$

for $k_p = 15$.

With $k_p = 15$ and $k_i = 0.3$ the system transfer function is

$$\frac{X}{U_c} = \frac{0.06z - 0.0588}{z^3 - 2.6z^2 + 2.264z - 0.6628} \tag{6.14}$$

which has poles at $z = 0.811 \pm j0.142$ and $z = 0.979$. The system open-loop bode plot is shown in Figure 6.12, which shows a phase margin of 60° at 0.5Hz and a gain margin of 15dB at 2Hz.

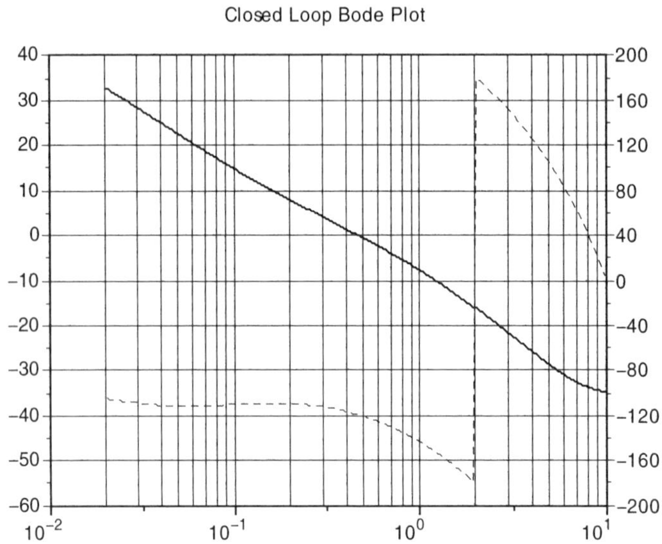

Figure 6.12 Pressure cuff system final bode plot

Systems with PI controllers as shown in Figure 6.9 almost always display a large overshoot, as a direct result of the zero that forms in the numerator of the transfer function, of which (6.14) is typical. This problem can be avoided, at the expense of a much longer settling time, by putting the proportional compensation in the feedback and the integral in the forward path, as in Figure 6.13. When this is done the transfer function becomes

$$\frac{X}{U_c} = \frac{0.0012}{z^3 - 2.6z^2 + 2.264z - 0.6628}. \tag{6.15}$$

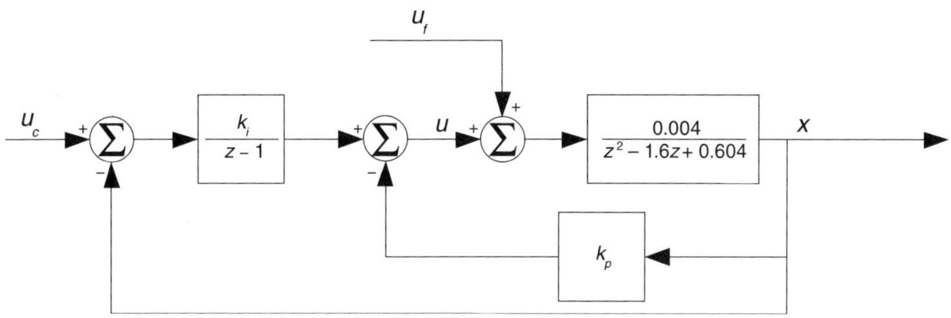

Figure 6.13 Plant with PI control

Figure 6.14 shows the time response of the system with the topology in Figure 6.9 (dotted line) and the topology in Figure 6.13. Either one of these responses may be the "better" one depending on the requirements of the product.[1] In addition you can always split the proportional gain between the forward and the feedback path; this will allow you to make a trade off between the amount of overshoot and the length of the rise time.

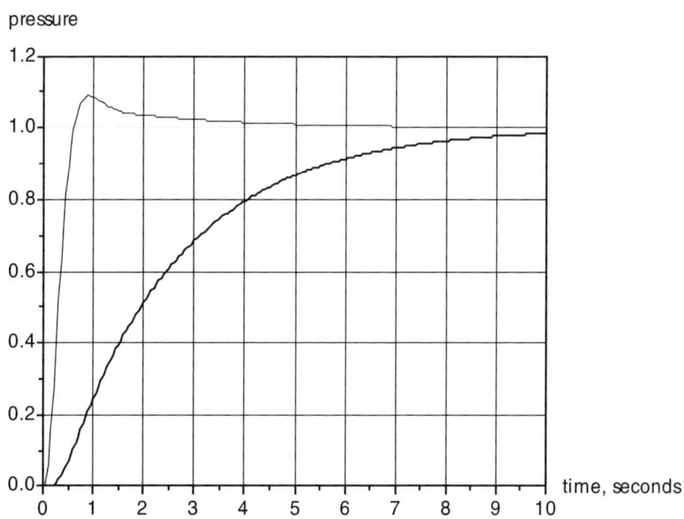

Figure 6.14 Time response of pressure cuff and controller

1 See also the section on integrator windup in Chapter 8.

Cascaded Integrators

I stated without proof that a system with integral control will have zero DC error. This statement needs to be made with precision, and extended to the benefits of a system with multiple integrators. Figure 6.15 shows a system with a number of integrators in the plant and a number of integrators in the controller. The system is shown with an intended input (u_i) as well as a disturbance input (u_d).

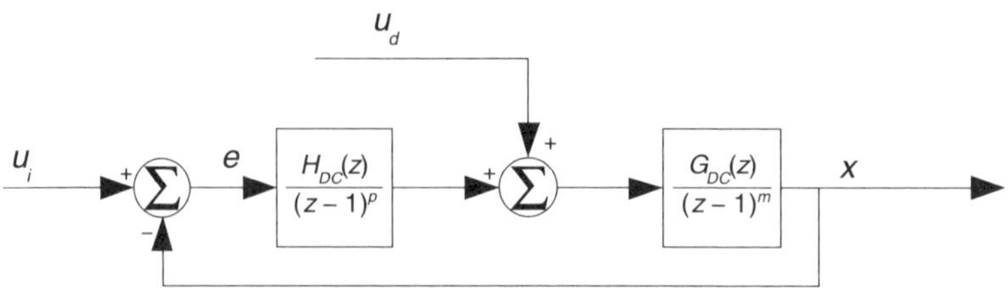

Figure 6.15 A loop with multiple integrators

To facilitate analysis, both the plant and the controller transfer functions are split into two parts: all of the integrators for the transfer function, and the rest of the transfer function. The integrators are lumped together into the denominator, with the controller having *k* integrators and the plant having *m* integrators. The balance of the transfer functions (subscripted "DC") have some finite, nonzero gain, and are shown in the numerators. It is assumed that the system is stable, with all closed loop poles having values $|z| < 1$.

The first property to look at is disturbance rejection. The transfer function from the disturbance input to the error signal, *e*, is

$$\frac{E}{U_d} = \frac{\dfrac{G_{DC}(z)}{(z-1)^m}}{1+\dfrac{G_{DC}(z)H_{DC}(z)}{(z-1)^{p+m}}} = \frac{(z-1)^p G_{DC}(z)}{(z-1)^{p+m}+G_{DC}(z)H_{DC}(z)} = \frac{(z-1)^p G_{DC}(z)}{D(z)} \tag{6.16}$$

where $D(z)$, the transfer function denominator, is the system polynomial.

From this we can deduce that if $p = 0$ and u_d is a step function, then the error signal will settle to a nonzero value. This is done by using the final value theorem. First, we find the z transform of the error signal:

$$E = \frac{(z-1)^0 G_{DC}(z)}{(z-1)^m + G_{DC}(z) H_{DC}(z)} \frac{1}{z-1}. \tag{6.17}$$

Next we find the final value of e using the final value theorem:

$$e_\infty = \lim_{z \to 1} (z-1)E = \frac{G_{DC}(z)}{D(z)} \frac{z-1}{z-1}. \tag{6.18}$$

By definition, $G_{DC}(z)$ is finite and nonzero at $z = 1$, and by definition, $D(z)$ is a polynomial in z with no zeros at $z = 1$ and is therefore finite and nonzero at $z = 1$. Because of these two nonzero terms, the leading ratio is nonzero. Because the two $z - 1$ terms cancel out, the final value of the error signal is nonzero.

Conversely, for any $p > 0$ the step response to a disturbance at u_d is zero. For such a case (6.18) becomes

$$e_\infty = \lim_{z \to 1} (z-1)E = \frac{G_{DC}(z)}{P(z)} \frac{(z-1)^{p+1}}{z-1}, \tag{6.19}$$

because the final ratio has an excess of $z - 1$ terms the limit evaluates to zero for all $p > 0$.

What can we say if the function u_d is a ramp, or parabola, or some higher-order function of time? In that case I will observe without proof that the z transform of a polynomial in time is

$$\mathbf{Z}\left(u(k)k^q\right) = \frac{Q(z)}{(z-1)^{q+1}}, \tag{6.20}$$

where $Q(z)$ is a polynomial in z with no zeros at $z = 1$ (recall that in this context $u(k)$ is the unit step function). Using this expression for u_d and solving for the final value we get

$$E = \frac{(z-1)^p G_{DC}(z)}{(z-1)^m + G_{DC}(z) H_{DC}(z)} \frac{Q(z)}{(z-1)^{q+1}} \tag{6.21}$$

and

$$e_\infty = \lim_{z \to 1} (z-1) E = \frac{G_{DC}(z) Q(z)}{D(z)} \frac{(z-1)^{p+1}}{(z-1)^{q+1}}. \tag{6.22}$$

From this we can see that if the number of integrators in the controllers, p, exceeds the order of the disturbance, q, then the error will be zero. If the numbers match then there will be some finite, constant error. Finally, if the order of the disturbance exceeds the number of integrators in the controller, then the error will be constantly growing.

Note that in the real world there is a distinct limit to all of this: an ever-growing disturbance to the plant implies that there needs to be an ever-growing control signal to counteract it. If left to go on forever, the controller would hit some limit, so any analysis of this type must take into account how long the disturbance will persist in reality, and what magnitude of control input must be maintained to counteract it.

In a very similar vein, we would like to know the capability of the system to follow some desired input, which may be a step, ramp, parabola or higher-order function. In this case, we want to look at the transfer function from the desired input to the output:

$$\frac{E}{U_i} = \frac{1}{1 + \frac{G_{DC}(z) H_{DC}(z)}{(z-1)^{p+m}}} = \frac{(z-1)^{p+m}}{D(z)}. \tag{6.23}$$

It should be easy to see that here the ability of the system to follow a higher-order input is a function not only of the number of integrators in the controller, but the number of integrators in the plant,[2] as well. The count of bare integrators, $p+m$, is called the *type* of the system, so a type 0 system has no integrators and cannot guarantee zero error to a step, a type I system has one integrator in either the plant or controller and has zero DC error to a step, but a finite DC error to a ramp, and so on.

Derivative

We have seen the use of proportional control, which takes a plant "as is" and implements control on a "take what you can get" basis. Integral control will add DC precision at the cost of reducing stability margins, but it cannot let us speed a system up. To increase the speed of a system, we use something that is the opposite of integral control—derivative control.

Where integral control adds DC precision and generally reduces stability, judicious use of derivative control can allow us to extend the loop closing frequency while keeping a system reasonably stable. We must use caution when designing with derivative control however—derivative action increases gain at high frequency, which raises problems with noise and with system stability at high frequencies that cannot be ignored in serious control system design.

When you use the Evans root locus plot to view the effect of adding derivative action to a controller, you see that a derivative adds a zero. This zero has the effect of pulling the root loci towards it; this can often have the effect of keeping the system stable at a higher gain (and frequency) than was achievable without the added zero.

Figure 6.16 shows an Evans root locus of a system consisting of a double integrator; the plot on the left shows the system without a differentiator, the plot on the right shows the system after cascading it with a PD controller whose transfer function is

$$H_{PD}(z) = 0.5 + 4.5\frac{z-1}{z}. \tag{6.24}$$

---

2 One should be careful, however, when counting integrators in the plant, to make sure that they are true integrators and not just low-pass filters whose deviation from an ideal integrator has been neglected for other analyses.

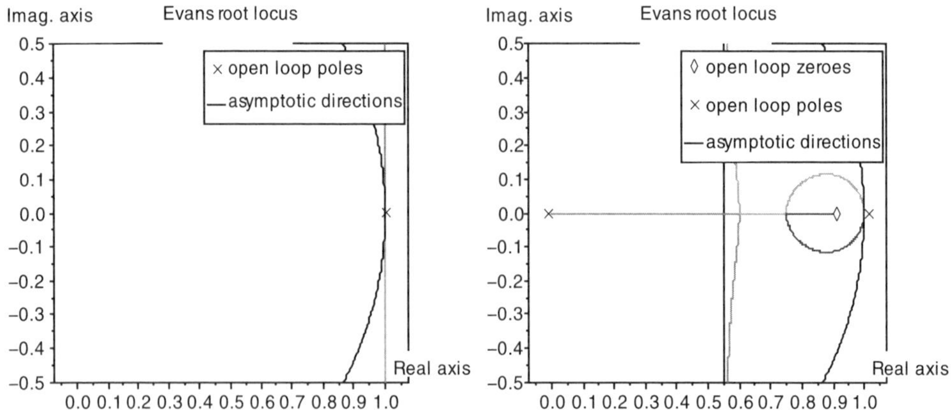

Figure 6.16 Derivative action with double integrator

A salient point to note about (6.24) is that in addition to adding a zero, it also adds a pole to the transfer function. This is an unavoidable feature of derivative action in a real system. This added pole means that while part of the root locus is pulled toward the zero, the total number of excess poles in the system stays the same—the difference is that the likely frequency at which the excess poles will exit the unit circle will be higher than without the differentiator.

Figure 6.17 shows the Bode plot of the differentiator with transfer function

$$H_D(z) = \frac{z-1}{z}. \tag{6.25}$$

It has several important properties that affect how well control can be carried out with a differentiator. The first, and most useful property is that the phase shift is positive 90°—this is what makes differentiators useful to us. The second property is that as the frequency goes towards half the sampling rate, the amount of positive phase shift diminishes down to 0, which means that as we try to push a loop's closing frequency up to any significant fraction of the sampling rate, we will get diminishing returns from our differentiator action. The final property is that as the frequency increases, the gain does as well. This last property means that we are increasing gain right where it can be the most inconvenient—not only is a system most likely to have less certain phase and gain characteristics at high frequencies, but if there is any measurement noise, it will be amplified to a disproportionate degree at these high frequencies. When this happens, the high level of high frequency noise can cause problems ranging from audible noise to burning actuators. Often it can be so severe that the drive to the actuator is overwhelmed by noise and any desirable control signal is drowned out, essentially rendering the control system useless.

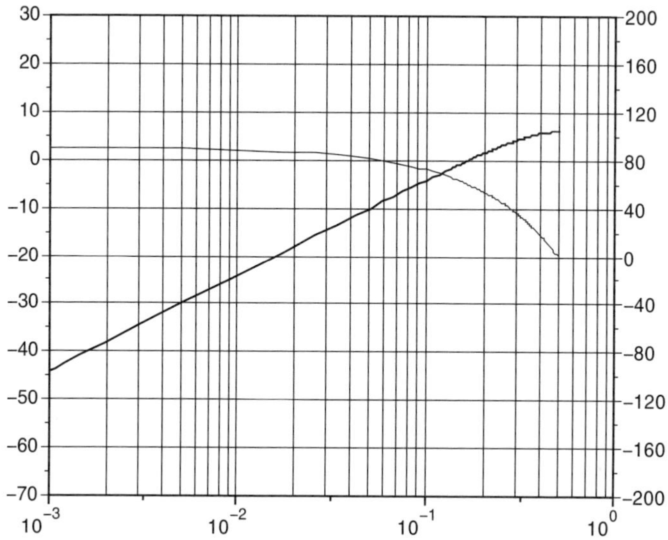

Figure 6.17 Bode plot of differentiator

Lead-Lag

Most of the problems with using a differentiator as shown in (6.25) can be overcome by limiting the upper frequency at which it acts. This is done by modifying the pole position:

$$H_D(z) = \frac{(1-d)(z-1)}{z-d}. \tag{6.26}$$

Now, instead of having a pole at $z = 0$, the differentiator pole is closer to 1. The differentiator will have the same characteristics at low frequencies as the one in (6.25), but its gain will level off under the control of the pole at $z = d$, just as if you had cascaded the differentiator with a first-order low-pass filter. This compensator goes by the names *lead-lag* compensator, damped differentiator and band limited differentiator.

As an example of how this may be used, assume that we must control the system shown in Figure 6.18. The system suffers from measurement noise y_n; it has a plant which has been modeled with the transfer function

$$G(z) = \frac{0.01}{z - 0.99}. \tag{6.27}$$

Instead of using a PID with a simple differentiator, we'll use a band limited differentiator:

$$H(z) = k_p + \frac{k_i}{z-1} + k_d \frac{(1-d)(z-1)}{z-d}. \tag{6.28}$$

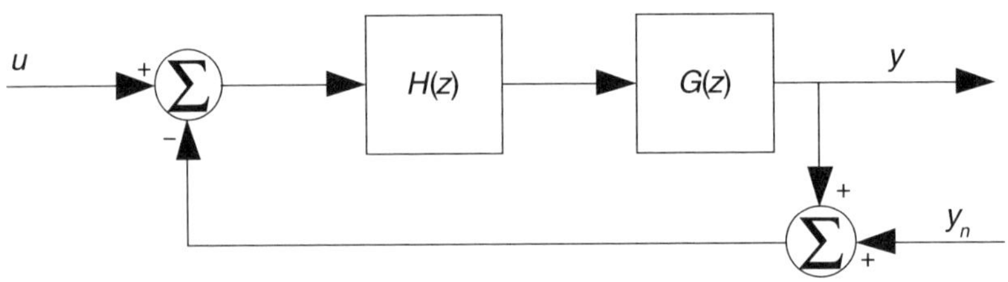

Figure 6.18 System with measurement noise

If we were to design a controller for this system with $k_p = 5$, $k_i = 0.06$, $k_d = 50$, and we left the differentiator undamped (that is, we let $d = 0$), then at high frequencies the open-loop gain levels off at around –5dB; should there be any process in the plant that causes the plant gain to rise above 5dB, then the system would go unstable. Of more immediate concern for this plant is that at high frequencies, the gain from the input noise to the plant drive exceeds 40dB—this could cause real problems when there's significant high-frequency noise.

On the other hand, we can keep the gains exactly the same and set $d = 0.9$ without affecting the system gain or phase margins to any great degree. What we *have* changed, however, is the gain at high frequencies: the open-loop gain goes to zero, and the noise-to-drive gain levels off slightly above 20dB, as shown in Figure 6.19 and Figure 6.20.

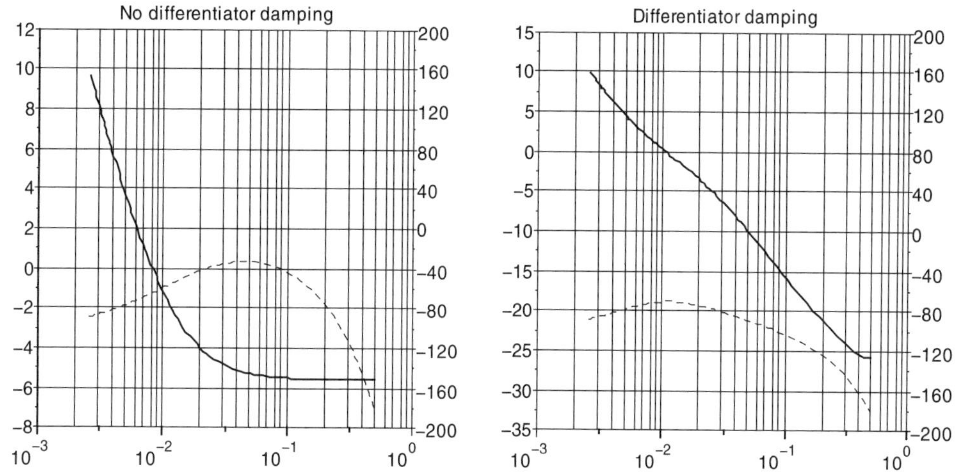

Figure 6.19 Loop gain with lead-lag compensator

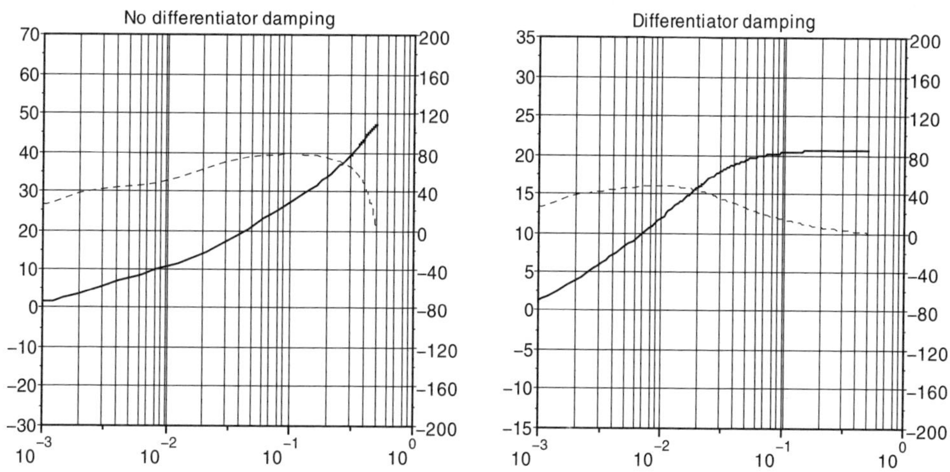

Figure 6.20 Noise coupling with a lead-lag compensator

Any time you are contemplating adding a derivative term to your controller, you should plan on using some damping in your differentiator; the only time not to use it is if the sample rate is low compared to the loop closure frequency, in which case there won't be much room for the differentiator's frequency response to get out of hand.

Lag (Low-Pass)

Lag compensation comes in one of two forms: either the system designer wishes to avoid problems caused by the infinite DC gain that come about when using an integrator, or the leveling-off of the gain in a lead-lag compensation scheme such as shown in Figure 6.20 is insufficient to suppress a very high noise level.

In the first case, you would want to implement a leaky integrator of the form

$$H_I(z) = \frac{1}{z-(1-\epsilon)} \tag{6.29}$$

which will act like an integrator for frequencies above ε radians/sample, but which levels off to a DC gain of $1/\varepsilon$ for very small frequencies. Much of the analysis of systems with integrators applies, except that there will be some residual steady-state errors to step, ramp and other inputs.

In the case of reducing noise, you would implement a low-pass filter:

$$H_{lp}(z) = \frac{1-d}{z-d}. \tag{6.30}$$

This filter has a DC gain of 1, and doesn't have a significant effect on the system until the frequency gets close to (1–d)/10 radians/sample. It *will* tend to destabilize the system at higher frequencies, so you must pay attention to its presence when you design it in, but if you're having noise problems, you can sometimes go a long way toward solving your problems with a simple lag filter.

Other Compensators

The compensators presented so far are not the only compensators out there, of course. Each control problem is unique, and sometimes unique compensators are demanded to get a system to behave correctly. If you have a plant that has some unusual characteristic, or if you have an unusual requirement that cannot be met by more conventional means, you should not shy away from using an oddball compensator just because it isn't described here—just make sure that you understand its effects and plan accordingly.

6.4 Design Flow

A common mistake in control system design is to assume that you have an adequate plant with fixed characteristics along with reachable performance goals, and that your job is just to design a controller. You may not have been the one to make this assumption, of course; often, control systems are thrown together with little understanding of the ultimate performance capabilities of the plant or controller hardware chosen.

Haphazard design like this is not the best road to good system performance. Where you can, you should try to get in on the design process as early as possible, and to influence the choices made in obtaining or designing the plant and its sensors. Poor choices in these areas can lead to results that range from inconvenient to disastrous. Where you cannot get in on the design process early, you should be prepared to engage in some clever tactics to make your system design succeed in spite of adversity—and you may find that your most important problem-solving skill is diplomacy rather than any technical knowledge.

Ideally, when you go to design a control system, you are given a set of performance specifications in a format that is easily related to the performance parameters in Chapter 4, some constraints about costing and specifications about the environment that the system will have to operate in, and then you (or your team) are given a free hand to design a system from the ground up.

Of course, this almost never happens in practice. Performance specifications may be vague, poorly stated, or nonexistent (at least until someone sees how your system works). Plant and sensor choices may be dictated by circumstances over which you have no influence. Often, you will be upgrading an existing product, and your task will be to use as much of the existing design as possible—this can add some severe constraints that can be either deeply rewarding or deeply frustrating to overcome.

In an ideal world, your design flow would go something like the steps below.

1. Specify the system performance and design constraints.
2. Investigate plant, sensor and controller options.
3. Design a candidate controller topology and choose a sampling rate.
4. Find initial tuning parameters for the controller from theory.

5. Prototype the controller, estimate software load, and choose an adequate processor.
6. Write the control software, test it on the actual plant, tune.
7. Ship product.

While you cannot follow this approach exactly in the real world, you should still come as close as you can. In particular, it is essential that someone thoroughly specify performance expectations in a manner that can be measured and designed to. Even if you are called (or stumble) into the project late in the game, you should be prepared to communicate clearly what the performance is going to be that you can achieve, and be ready to make the system perform to that level.

6.5 Conclusion

The process of actually designing a control system and getting it working is probably the most rewarding part of embedded control system work. This chapter has covered the essential bits and pieces that you can use in designing control systems.

7

Sampling Theory

So far, the discussion has been about the behavior of systems from the perspective of the processor or a piece of software. Time has been treated as happening in discrete steps, with nothing happening in between. In the real world, this is not the case; in the real world, time is a continuous process. This introduces three questions that we must answer before we can interface a digital controller to the real world: how do we get feedback data into the controller, how do we get command data out of the controller, and how do we model the real-world processes that we wish to control?

These questions are answered by sampling theory and the Laplace transform. This chapter presents an abbreviated take on sampling theory, with emphasis on those areas most important to control systems design. It presents different ways that a signal may be sampled, and gives the reader tools for determining system behavior given different methods of sampling.

7.1 Sampling

Sampling is the process by which continuous-time quantities are turned into a sequence of discrete-time measurements. Most control systems will convert the appropriate plant measurements to voltages and apply them to an analog-to-digital converter (ADC) that will sample the voltage and convert it.[1] Some measurements may be done in other ways, such as with a rotary encoder coupled with a counter, but such a system must still be sampled at a discrete point in time for the controller to use the data. However it is done, at some point the measurement is converted from continuous time to discrete time.

---

1 Older ADCs and some very specialized systems may require external sampling circuitry, but nearly all current ADC integrated circuits include their own sampling circuitry on board.

Figure 7.1 shows how a sampler may appear in a block diagram. It is simply a picture of a switch, with the sampling interval indicated underneath the switch (the 'T' in the figure). If the sampling is synchronous with some other event, this will be denoted by replacing the arcing arrow with a dotted line to the synchronizing event.

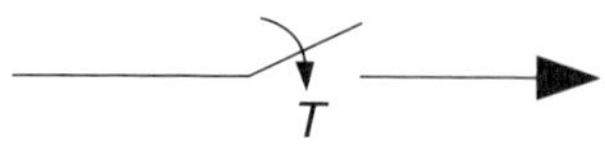

Figure 7.1 Showing sampling in a block diagram

Sampling is easy to represent mathematically: given a signal x to be sampled, the sampled version of x is simply x taken at those sample times:

$$x_k = x(Tk). \tag{7.1}$$

Figure 7.2 shows the result of sampling a signal. The upper trace is the continuous-time signal, while the lower trace shows the signal after sampling once per millisecond. Note that the lower trace shows no signal between samples. This is because after sampling there *is* no signal between samples—any information

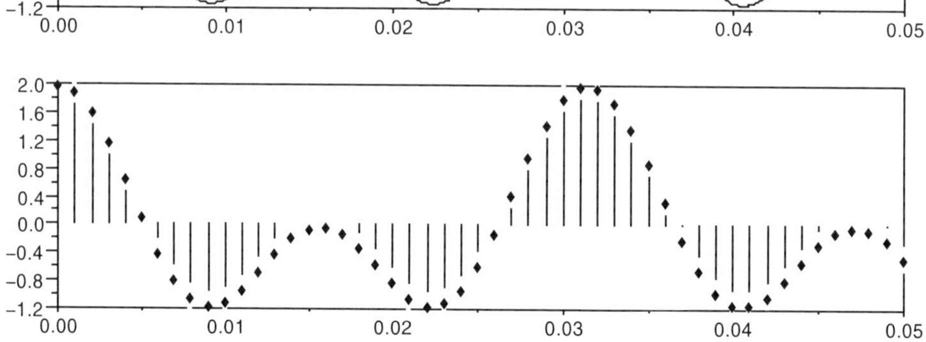

Figure 7.2 The results of sampling

that may have existed between the samples in the original signal is irretrievably lost in the sampling process.

7.2 Aliasing

By ignoring anything that goes on between samples, the sampling process throws away information about the original signal. This information loss must be taken into account during system design. The effect is called *aliasing*, and it is easily expressed and modeled as a frequency-domain phenomenon.

Take a signal that is a pure sinusoid, and look at its sampled version:

$$x_k = \cos(\omega Tk). \tag{7.2}$$

If you know the frequency of the original signal, you'll be able to predict the sampled signal exactly. But the sampled signal won't necessarily be at the same frequency as the original signal: there is an ambiguity in the signal frequency equal to the sampling rate. This can be seen if you consider two signals, one at frequency f and one at frequency $f + 1/T$. The sampled version of these two signals will be exactly the same:

$$\cos\left(2\pi\left(f + 1/T\right)Tk\right) = \cos\left(2\pi k + 2\pi fTk\right) = \cos\left(2\pi fTk\right) \tag{7.3}$$

This ambiguity is aliasing.

Figure 7.3 shows an example of aliasing. Two possible input sine waves are shown: one has a frequency of 110Hz, the other has a frequency of 1110Hz. They are sampled at 1000Hz. The dots show the value of the sine waves at the sampling instants. As indicated by (7.3), these two possible inputs both result in *exactly* the same output.

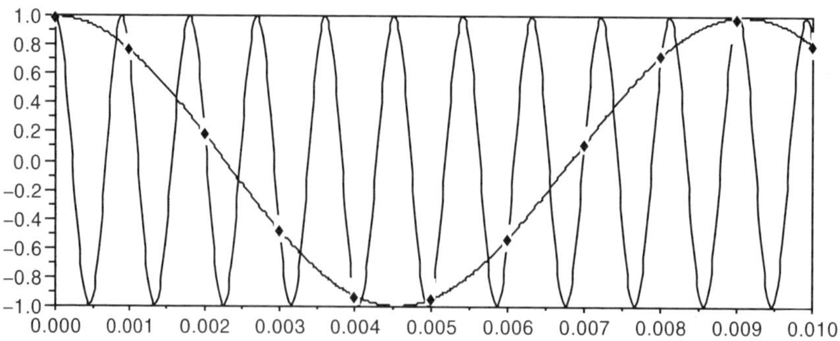

Figure 7.3 Aliasing of two sine waves

Continuous-time signals can be thought of as consisting of a sum of different signals at different frequencies.[2] This distribution of a signal's energy over frequency can be shown as a plot of spectral density vs. frequency. Aliasing will make these signals indistinguishable from signals spaced an integer number of sampling rates away.

Figure 7.4 shows this effect. The central frequency density plot is the signal that's being sampled; the others are the signal's aliases in sampled time. The signal energy appears to "fold back" at half the sampling rate.

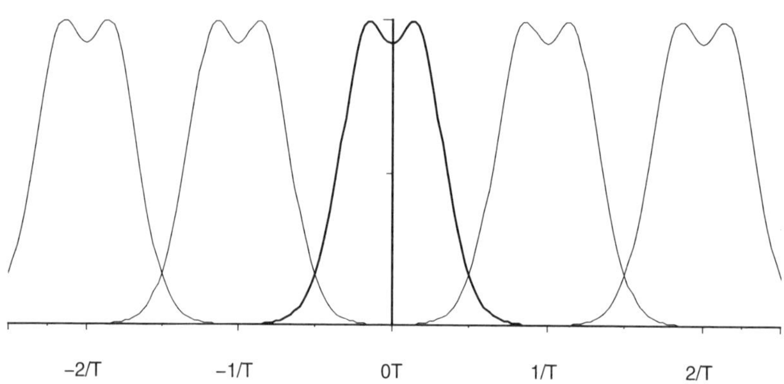

Figure 7.4 Aliasing of a signal spectrum

[2] This can be expressed formally using the Fourier transform; I will restrict myself to an informal discussion here.

Aliasing is of central concern for systems that must accurately reconstruct a signal whose bandwidth is a large proportion of the sampling rate, and for which increasing the sampling rate is not possible or economical. Communications systems, particularly digital video systems, can have very critical requirements related to the suppression of aliasing. In control systems, however, sampling rates are usually much larger than the bandwidth of the expected plant motion. Even when they are not, one doesn't build a control system for accurate reconstruction of a signal, only for good control. Consequently, aliasing is not a direct concern unless you have reason to expect that noise will be injected into your feedback signal before it is sampled. Only when the feedback contains high-frequency noise components that may be aliased to within the bandwidth of the control system should you need to worry about aliasing.

If you do contemplate adding anti-alias filtering before your ADC conversion, pay close attention to the phase shift of the filter—filters with sufficient rolloff to be useful for anti-aliasing have, almost by definition, high phase shifts at surprisingly low frequencies. Anti-aliasing filters that are suitable for control systems must be carefully designed not to reduce the phase margin in the system's intended bandwidth while still attenuating undesired signals.

7.3 Reconstruction

Because a sampled-time signal isn't defined between samples, it can't be used directly in a continuous-time system. To use a sampled-time signal in a continuous-time system, it must be converted into a continuous-time signal before it can be useful. This conversion from sampled to continuous time is called *reconstruction.*

Reconstruction is done by interpolating the output signal. Interpolation is the process whereby the value of the continuous-time signals between samples is constructed from previous values of the discrete-time signal. In discrete-time control systems, this interpolation is almost universally done with digital-to-analog circuits, which contain an intrinsic zero-order hold.

A zero-order hold is simply a device which takes on the value of the most recent sample and holds it until the next sample time:

$$y(t) = y_{floor(t/T)}, \qquad (7.4)$$

where "floor" is simply the floor function as you might find in the C math library.

In a block diagram a zero-order hold is indicated by a block containing a picture of an interpolated signal, as shown in Figure 7.5. This is exactly the function provided by most digital-to-analog converters: the processor writes a number to the DAC, which drives a corresponding voltage (or current) output until the next number is written.

The result of using a zero-order hold is shown in Figure 7.6. The sampled-time signal is at the top, and the interpolated signal is on the bottom. Note the

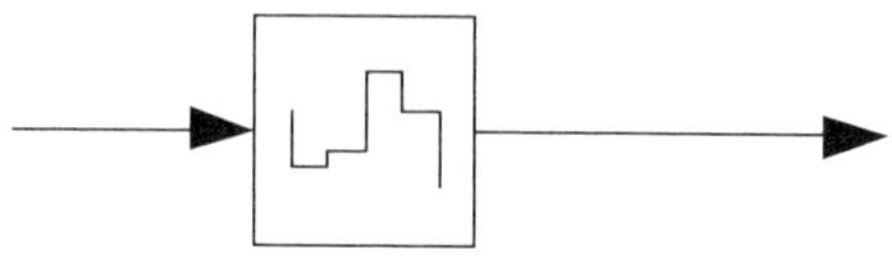

Figure 7.5 A zero-order hold

"stair-step" nature of the interpolated signal, as well as the fact that on average it is delayed from the original signal—this bit of delay is an inevitable part of implementing controllers in sampled time.

Reconstruction causes a related phenomenon to aliasing:[3] when a signal at a certain frequency is reconstructed, the continuous-time signal will have components at the sampled-time signal frequency, as well as all multiples of the sample rate plus and minus the sampled-time signal frequency. For example, if the sampled-time signal is

$$y_k = \cos\left(2\pi f_0 k\right) \tag{7.5}$$

then the reconstructed signal will have components at

$$f = f_{0,} \frac{1}{T_s} + f_{0,} \frac{1}{T_s} - f_{0,} \frac{2}{T_s} + f_{0,} \frac{2}{T_s} - f_{0,} \ldots . \tag{7.6}$$

[3] One can unify the sampling and reconstruction processes with careful use of the Fourier transform. Such analysis can be a useful tool, but would require a lengthy digression from the core content of this book.

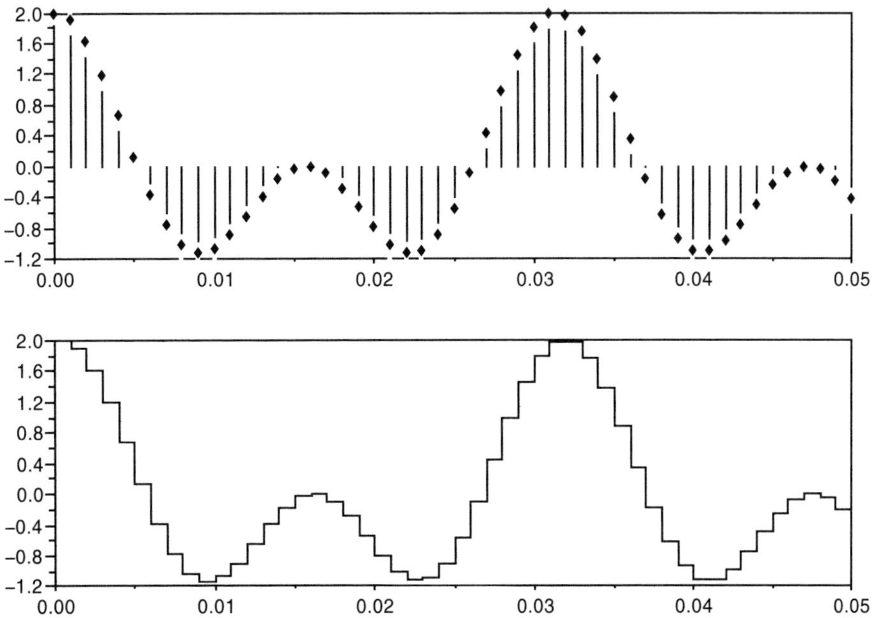

Figure 7.6 Reconstruction

The zero-order hold interpolator acts as a low-pass filter for the signals going through it, with a frequency amplitude response of

$$A = sinc\left(\pi T_s f\right) = \begin{cases} 1 & f = 0 \\ \dfrac{\sin\left(\pi T_s f\right)}{\pi T_s f} & otherwise \end{cases}. \tag{7.7}$$

This means that while all of the components shown in (7.6) will be generated, the high-frequency components will be attenuated before reaching the plant.

In audio and video circuits, these high-frequency components, as well as the rolloff of the low-pass filter described by (7.7), can be of great concern. In such cases, system designers may go to great lengths to design reconstruction filters to filter their DAC output signals. As with aliasing, however, in a control system this added filtering is usually unnecessary and often adds phase shifts that are detrimental to control system performance. In general, the system's plant will filter out any high-frequency components; if such filtering is inadequate, or if the simple zero-order-hold action creates other noise problems, a low-pass filter with a high cutoff frequency can be used to knock the corners off the DAC output before it is amplified and applied to the plant.

7.4 Orthogonal Signals and Power

Any two signals, whether random or not, have some degree of linear dependence. If the two signals are completely dependent then one signal is simply a multiple of another; for example, if x and y are perfectly dependent then

$$y_k = Ax_k \,. \tag{7.8}$$

In such a case their sum is easily found:

$$x_k + y_k = \left(A+1\right)x_k \,. \tag{7.9}$$

In signal processing theory, the degree to which two signals are not linearly dependent is called their *orthoganality*. Signals are orthogonal over some interval if the average of their product over that interval is zero; for example, if x and y are orthogonal over the interval $[k_1, k_2]$ then

$$\frac{1}{k_2 - k_1}\sum_{k=k_1}^{k_2} x_k y_k = 0 \,. \tag{7.10}$$

This interval can be extended to infinity, so

$$\lim_{N\to\infty}\frac{1}{N}\sum_{k=0}^{N} x_k y_k = 0 \tag{7.11}$$

indicates that x and y are orthogonal.

For any two orthogonal signals, their powers[4] will add. For example, if you have a signal that is the sum of the signals x and y from (7.11),

$$w_k = x_k + y_k \,, \tag{7.12}$$

[4] In the world of signal processing, engineers can play fast and loose with the term "power." In this context it simply means the average of the square of a signal. Of course, in control systems "power" also can mean "that which makes my actuators overheat and my batteries drain"—it is a good idea to remember which meaning you are dealing with at any given time.

then the power of the resulting signal is simply the sum of the powers of the two components:

$$\lim_{N\to\infty}\frac{1}{N}\sum_{k=0}^{N} w_k^2 = \lim_{N\to\infty}\frac{1}{N}\sum_{k=0}^{N}\left(x_k + y_k\right)^2 = \lim_{N\to\infty}\frac{1}{N}\sum_{k=0}^{N} x_k^2 + \lim_{N\to\infty}\frac{1}{N}\sum_{k=0}^{N} y_k^2 . \quad (7.13)$$

If, however, you have a pair of signals that are perfectly correlated, as in (7.8), the resulting power will depend on the correlation:

$$\lim_{N\to\infty}\frac{1}{N}\sum_{k=0}^{N} w_k^2 = \lim_{N\to\infty}\frac{1}{N}\sum_{k=0}^{N}\left(x_k + y_k\right)^2 = \lim_{N\to\infty}\frac{1}{N}\sum_{k=0}^{N}\left[\left(A+1\right)x_k\right]^2 = \lim_{N\to\infty}\frac{\left(A+1\right)^2}{N}\sum_{k=0}^{N} x_k^2 . \quad (7.14)$$

Some real-world examples of correlated signals are the input and output of an amplifier, or the command and response of a servo-loop. Uncorrelated signal pairs include sinusoidal signals at two different frequencies, or two sinusoidal signals at the same frequency but exactly 90 degrees out of phase, and signals that include noise.

Signals can be partially correlated as well, with one signal being partially dependent on the other, but including an uncorrelated component.

7.5 Random Noise

Many processes that affect control system operation take on the appearance of random signals that interfere with the desired operation of the system. These signals may not actually be random, but it is often convenient to treat them as if they were for the purposes of control system design.

You cannot talk about random signals in exactly the same way that you can talk about ordinary, deterministic signals. A deterministic signal is known: it is a step, a ramp, a sinusoid or some other predefined thing. A random signal, on the other hand, is unknown before you measure it. Even after you've measured part of a random signal, you still won't be able to predict exactly what it will be for subsequent measurements. Nonetheless, a random signal can have characteristics that distinguish it from other signals, and this can be used to make partial predictions about its behavior.

This difficulty that we have with talking about random signals has, indeed, led to a distinction that we make when talking about them, which is to separate a noise *signal* from the noise *process* that generates it. Take the example of flipping a coin. If we assigned 1 to heads and 0 to tails, then the results of a particular session of coin flipping may go something like $x_k = \{1, 0, 0, 1, 1, 0, 0, 1, 0, 0 ...\}$. In this case, the string of resulting 0's and 1's is the random signal;the act of flipping the coin is the generating process.

We characterize random processes by a number of different measures.

- A probability distribution, such as Gaussian, uniform or discrete-value (such as our coin-tossing experiment, which can take on heads or tails).
- A mean value; that is, the expected average value of the signal generated by the random process. In the case of our coin-toss, this is exactly half for a fair coin.
- A variance;[5] that is, the expected value of the square of the amount that the signal varies from its mean value. Note that the power of a random signal is equal to its variance plus the square of its mean value.
- A spectral density, which describes how the noise power is distributed in frequency and can be used as a measure of how much one sample in the random signal can be used to predict the value of another.

If a random variable is Gaussian, then it can be completely characterized by its mean and variance. Similarly, if a random process is Gaussian, then it can be completely characterized by its spectral density (which will tell you mean and variance).

A random signal can be said to have a certain noise power—this noise power is the average of the power that you would expect to measure on the signal. For systems in the digital domain, this power will usually be expressed as a percentage of full scale. Noise is often expressed as a root mean square (RMS) value. This is the square root of the noise power, and in these systems will often also be expressed as a percentage of full scale.

If you have two noise processes that are uncorrellated, then their respective noise powers will add; if one noise process generates noise at 9% of full scale and another one generates noise at 16% of full scale, then the sum of those two noise signals will have a noise power of 25% of full scale. If you view the same

---

5 In rare cases, it is possible for a random process to generate a signal with infinite variance. Most random signals you will encounter, however, will have finite variance.

situation in terms of RMS values, the first signal would have a level of 30% RMS, the second would have a level of 40% RMS, and the result would have a level of 50% RMS. Note that any time two uncorrelated signals are added, their RMS value is equal to the root of the sum of the squares of the RMS values of the two signals.

7.6 Nonideal Sampling

Note that Figure 7.2 and (7.1) show the result of ideal sampling: the sampled signal is assumed to be the exact value of the original signal at exactly the sampling point and none other. In real systems, this is never the case; the sampling can never be quite ideal. Depending on the nature of the system and its requirements, the impact of the nonideal sampling can range from totally insignificant to something that will dominate the engineering effort.

There are a number of effects that can affect sampling. Most of these effects can be lumped into three categories: effects that add noise to the signal, effects that act to filter the signal, and effects that act to add sampling jitter. This section presents an overview of the effects you may see; consult a good book on electronic circuits for more information on analog-to-digital conversion.[6]

Noisy Measurement

Noise can creep into a measurement in a number of ways. Signal conditioning electronics can pick up ambient electromagnetic interference, which is then amplified along with the intended signal and applied to an ADC, sensors can be subject to mechanical vibration or other nonelectrical noise, and the electronics (including the ADC) can inject their own noise into the measurement.

Any noise that finds its way into a measurement will be filtered by the system's response to input at that point and will show itself as error at the system output. Measurement noise can be particularly troublesome, because the system's response to measurements tends to be very close to its response to commands; reducing response to measurement noise means slowing down the system's response to commands.

---

6 Such as *The Art of Electronics* by Paul Horowitz and Winfield Hill.

The distribution of the noise in frequency and the place in the system where it occurs has a strong effect on what you may be able to do with it. You can view noise as being injected either before or after the sampling point, as shown in Figure 7.7. If you have noise that is internal to the system (n_i in the figure) you are stuck with it; it can only be dealt with by filtering or by varying the tuning of the system. Noise that is external to the system (n_e in the figure) can, if you are lucky, be reduced or eliminated before the sampler.

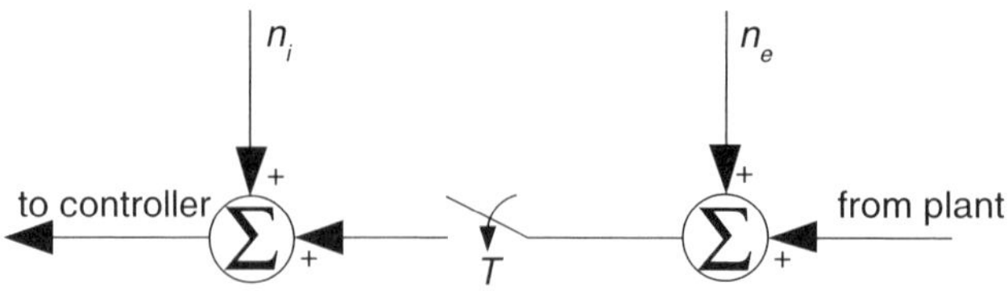

Figure 7.7 Noise injection points

Once noise that is within the intended bandwidth of the control system reaches the ADC, there is not much that the control engineer can do with it. Generally such noise is indistinguishable from actual plant noise;[7] if the level of noise causes unacceptable plant errors, then it must be dealt with at the source.

In-band noise can arise within the plant itself, in the sensor used to measure the plant output, in the electronics used to condition the sensor output, or it can be coupled into the sensor signals from external sources. Noise from the plant or from the sensor must be dealt with by changing the plant, or possibly by changing the sensor to measure some other aspect of the plant's response. Noise arising from the electronics must be dealt with by careful circuit design. If the noise is being generated somewhere else and is being coupled into the sensor signal, then this coupling must be found and reduced or eliminated.

Noise that is converted to a frequency that is outside of the intended control system bandwidth can be inconsequential, or it can be a serious problem. In this case, you not only need to be concerned with the effect of the noise on the plant output, but you also need to be concerned with the effect of the noise on the

[7] In special cases, some types of noise can be dealt with by nonlinear signal processing, but doing so is beyond the scope of this book.

plant drive: this drive level may be high enough to saturate the drivers, changing their characteristics, it may cause heating either in the driver electronics or in the plant itself, it may cause excessive power consumption when the system is supposed to be at rest, and it may cause undesirable secondary effects such as buzzing or vibrations.

Example 7.1 Dealing with Noisy Measurement

A motion-control system samples at 1kHz, with a closed-loop bandwidth of 4Hz. The plant transfer function in the z domain is

$$G(z) = \frac{0.0001z}{(z-0.99)(z-1)} \tag{7.15}$$

and the controller transfer function is

$$H(z) = 10 + (0.05)\frac{z}{z-1} + (500)\frac{z-1}{z}. \tag{7.16}$$

The system's position measurement is corrupted by noise that is Gaussian and white in the digital domain, with an RMS noise level of 0.1% of full scale. The system will malfunction if the noise at the output of the controller exceeds 50% of full scale for more than 20% of the time.

Find the RMS noise level of the system output, and at the output of the controller. Find the percentage of time that the controller's output exceeds 50% of full scale due to noise, assuming that the average controller output is zero. Find a method of reducing the noise at the controller output, and assess the impact of the system changes to the system's performance.

The system's closed-loop transfer function from the command (and measured) position to the output is given in Figure 7.8. This is a fairly normal plot, with some peaking before the amplitude dropoff, and a 3dB bandwidth of about 10Hz.

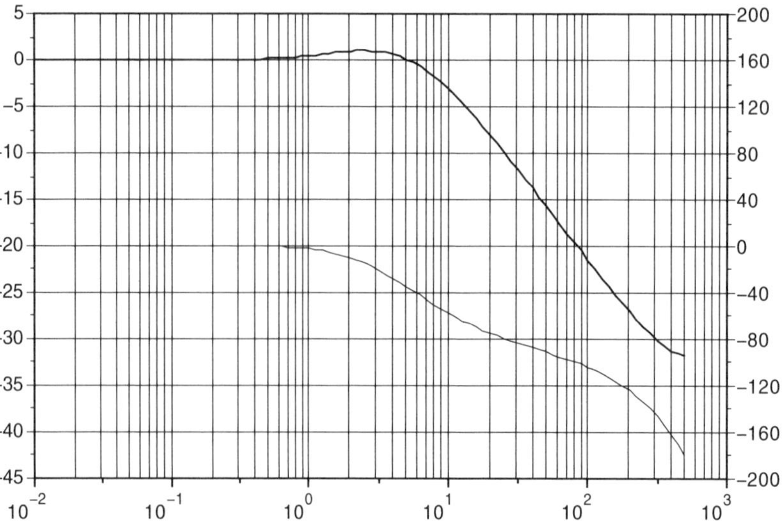

Figure 7.8 System response bode plot

Figure 7.9 shows the transfer function from the measured position to the drive output of the controller. Note that the gain from measurement to position exceeds 60dB at half the sampling rate—any high-frequency noise at the measurement will be amplified by nearly 1000 before it is applied to the output!

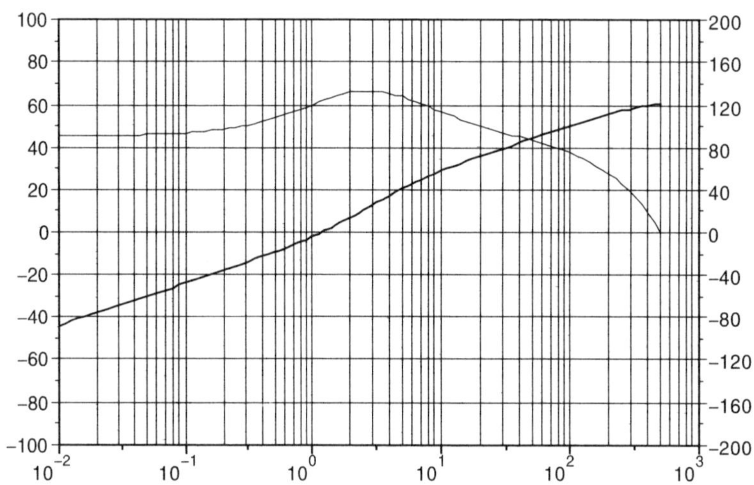

Figure 7.9 Bode plot of system output to position measure

To find the effect of the noise on the system output, we first need to find the noise spectral density at the output, and then we need to integrate this spectral density to find the actual noise power. The noise spectral density at the input can be found from the noise power, the fact that the noise is white and the sampling rate:

$$\int_0^{500Hz} \eta_0^2 df = 0.001^2 \Rightarrow \eta_0 = \frac{0.001}{\sqrt{500Hz}} = 44.7 \cdot 10^{-6} \frac{1}{\sqrt{Hz}}. \tag{7.17}$$

The noise density at the controller output is the above figure times the system gain from measurement to output, or

$$N_{RMS} = \eta_0 \sqrt{\int_0^{500\text{Hz}} \left| T_m \left(e^{-i\frac{\pi f}{500\text{Hz}}} \right) \right|^2 df} \tag{7.18}$$

where T_m is the transfer function from measurement to drive:

$$T_m(z) = \frac{H(z)}{1 + H(z)G(z)}. \tag{7.19}$$

This integral evaluates to N_{RMS} = 73% of full scale; with this level of noise, the drive will exceed 50% over 33% of the time.

If the derivative function in (7.16) is bandlimited, then the controller transfer function becomes

$$H(z) = 10 + (0.05)\frac{z}{z-1} + (500)\frac{(1-d_d)(z-1)}{z-d_d}. \tag{7.20}$$

Setting d_d to 0.85 does not seriously impact the controller's robustness or the response of the plant, as seen in Figure 7.10 (compare to Figure 7.8), but the response of the drive to measurement noise is down by over 20dB, as seen in Figure 7.11. Recalculating the drive noise gives N_{RMS} = 9%, with virtually no chance that the drive will exceed 50% as a result of measurement noise.

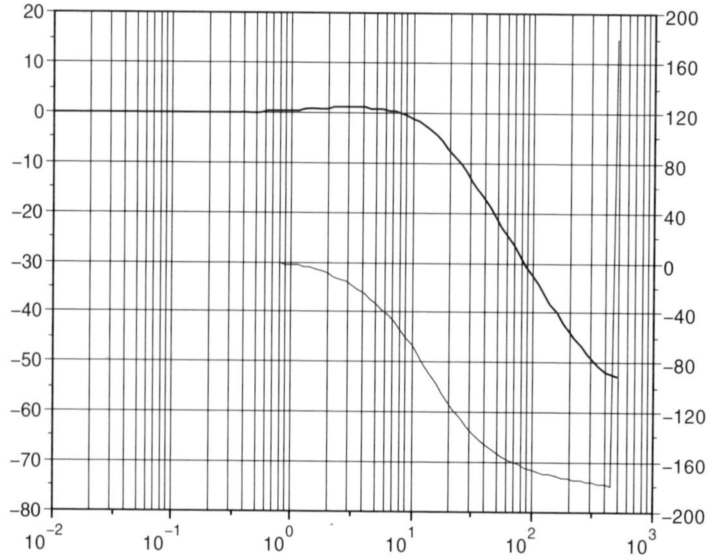

Figure 7.10 The system transfer function, with derivative filtering

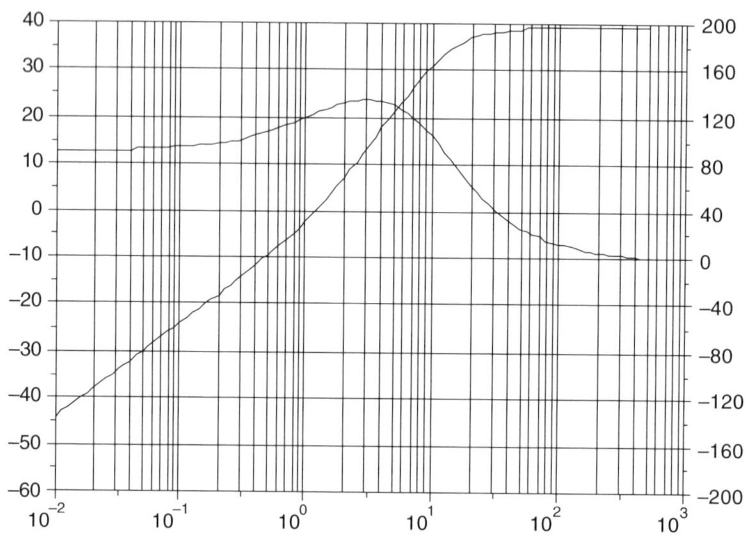

Figure 7.11 The transfer function from measurement to drive, with derivative filtering

Filtering

Any ADC will introduce some amount of frequency shaping and delay into a signal. This can range from insignificant (in a flash converter) to heavy low-pass filtering with significant delay (in a sigma-delta converter). Table 7.1 lists some popular ADC types, the amount of filtering you can expect to see, and their expected delay. With the exception of the sigma-delta type, most of these ADCs can be treated as a simple zero- or one-sample delay—but it never hurts to analyze a manufacturer's published information to determine just how much filtering and delay there is, and decide if the device is right for you.

ADC Type	Filtering Effects	Delay
Flash	Little to none	Insignificant
Successive Approximation	Very little, but some straight delay	Usually one sample time (more in a high-speed pipelined ADC)
Single-slope	Some low-pass filtering	Effectively zero to half a sample time
Dual-Slope	Comb at some frequency—commonly has zeros at 60Hz[8] and all its harmonics.	Effectively between half and one sample time
Voltage-to-Frequency	Depends on frequency to digital conversion	Depends on frequency to digital conversion
Sigma-Delta	Low-pass, filter shape varies with intended use	Multiple sample times, varies with intended use

Table 7.1 ADC types and filtering

8 Or whatever the local power-line frequency may be.

Sampling Jitter

Of these categories, sampling jitter is the only one that is unique to the sampling process. Sampling jitter occurs when the actual sampling time does not fall exactly on the sampling interval. When this happens (7.1) becomes

$$x_k = x(Tk + \epsilon_k) \tag{7.21}$$

where ε_k is the sampling time error for the kth sample; in general ε_k will be a random process, where each sample is unknown but the behavior of any set of values can be described statistically.

By itself (7.21) cannot be easily modeled within the framework of the z transform. We can apply some useful approximations to (7.21), however, that—used with care—can allow us to model the effects of jitter.

The Taylor's series expansion of (7.21) gives an exact expression:

$$x_k = x(Tk + \epsilon_k) = x(Tk) + \epsilon_k \frac{d}{dt} x(t)\bigg|_{t=Tk} + \frac{\epsilon_k^2}{2} \frac{d^2}{dt^2} x(t)\bigg|_{t=Tk} + \cdots. \tag{7.22}$$

If you assume that the double and higher derivatives of $x(t)$ aren't significant, this reduces to the approximation

$$x_k \simeq x(Tk) + \epsilon_k \frac{d}{dt} x(t)\bigg|_{t=Tk}. \tag{7.23}$$

This approximation can be further boiled down to

$$x_k \simeq x(Tk) + n_k, \tag{7.24}$$

where technically n_k depends not only on ε_k but also on the time derivative of the signal at the sampling instant. Combining (7.23) and (7.24) with some common sense will give us some useful results, however.

From (7.23) and (7.24), we can see that the degree to which sampling jitter can be a problem will depend on the rate of change of the signal to be sampled multiplied by the amount of jitter. In addition, the nature of the controller will have a strong affect on the severity of a control system's reaction to sampling jitter, with problems ranging from dire to insignificant.

A fast-moving plant coupled with a controller that has a high gain to instantaneous changes in the input (such as a plant with high derivative gain) will reflect input jitter into a large noise in the output. This output noise can easily be enough to overwhelm the desired control signal in the output and render the control system useless unless the sampling jitter is addressed.

On the other hand, a plant that moves relatively slowly, coupled with a controller that has a low gain to instantaneous changes in input (such as a plant with low proportional gain and a great deal of integral action) may not couple sampling jitter to the output to any significant degree whatsoever. If this is the case, then large amounts of sampling jitter can be safely ignored.

Example 7.2 A Controller with Jitter

A controller is to be designed and built with the sampling instants determined entirely by the action of software. The controller samples at 50Hz and implements a PID controller with the equation

$$H(z) = k_p + \frac{k_i z}{z-1} + k_d \frac{(1-d)(z-1)}{z-d}. \tag{7.25}$$

The controller is to be used to control two different plants.

The first plant is a heater unit which will require proportional-integral control with $k_p = 10$ and $k_i = 0.01$; at full power the heater unit heats the plant up at a rate of 1% full scale per control iteration. Due to the nature of the heater unit, the system can tolerate as much as a 10% variation from sample-to-sample in the output while the heater is warming up.

The second plant is a motor which will require proportional-integral-derivative control with $k_p = 10$, $k_i = 0.01$, $k_d = 2000$ and $d = 0.5$. At full power, the motor can travel as much as 10% per control iteration. Due to the nature of the mechanism, the system can tolerate no more than a 10% variation from sample to sample in the output as the motor is moving from one position to the next.

Determine the allowable amount of sampling jitter for each system.

The output noise resulting from sampling jitter is dominated by the controller's initial response to a change, which can be found by finding the first element in the transfer function's impulse response. For the controller in (7.25) this is

$$h_0 = k_p + k_i + (1-d)k_d . \tag{7.26}$$

For the heater plant, then, the initial response to any jitter is

$$y_k = (10+0.01)n_k \simeq 10 \in_k \frac{d}{dt}x(t)\bigg|_{t=Tk} . \tag{7.27}$$

If the derivative never exceeds 1% of full scale per sample, the maximum output noise will be

$$y_{max} = 10\left(\frac{0.01}{20ms}\right)\in_{max} . \tag{7.28}$$

To hold this maximum output noise to 10% of full scale we need to satisfy:

$$10\left(\frac{0.01}{20ms}\right)\in_{max} < 0.1 \tag{7.29}$$

which works out to

$$\in_{max} < 20ms . \tag{7.30}$$

Since this limit is just the sampling interval, clearly sampling jitter is not our most critical problem in this control system.

The analysis is similar for the motor, however in this case we have

$$y_k = (10+0.01+2000(1-0.5))n_k \simeq 10 \in_k \frac{d}{dt}x(t)\bigg|_{t=Tk} . \tag{7.31}$$

This becomes

$$y_{max} = 1000\left(\frac{0.1}{20ms}\right)\epsilon_{max} \tag{7.32}$$

and

$$1000\left(\frac{0.1}{20ms}\right)\epsilon_{max} < 0.1 \tag{7.33}$$

or

$$\epsilon_{max} < 20\mu s\ . \tag{7.34}$$

This figure is still reachable for most modern processors, but care must be taken in the software design to ensure that there isn't excessive jitter in the software timing.

Problems with sampling jitter are not limited to the input—some plant/controller combinations are sensitive to output jitter. A particular case of this is a highly resonant plant whose resonant peak is compensated by a notch filter. In such a case, the output-side jitter can act to spread the effective response of the controller's notch to a totally unacceptable degree.

Output jitter can be dealt with by clocking the system DACs with a hardware signal, or by taking measures in software to ensure a consistent time interval between the sampling time and the moment that the DAC output becomes effective.

7.7 The Laplace Transform

The Laplace and the z Transform

Consider the system shown in Figure 7.12. A sampled-time signal, u_k, is applied to a zero-order hold, and the result is applied to a continuous-time system. The continuous-time system acts on the signal, and its output, $y(t)$ is sampled. Is there a way to express how the continuous-time system behaves?

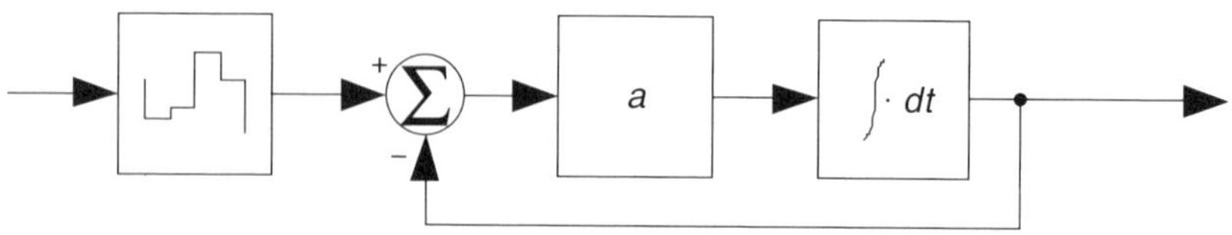

Figure 7.12 A mixed-time system

Looking at the continuous-time system you can see that it implements a differential equation. Specifically

$$\frac{d}{dt} y = a\left(u - y\right). \tag{7.35}$$

If you were to solve this differential equation for a unit step input, you would find that

$$y\left(t\right) = 1 - e^{-at}. \tag{7.36}$$

We can approximate the differential equation in (7.35) with a difference equation:

$$\frac{y\left(Tk\right) - y\left(T\left(k-1\right)\right)}{T} \simeq a\left(u\left(Tk\right) - y\left(T\left(k-1\right)\right)\right). \tag{7.37}$$

This is called a "backwards difference" equation, because it uses the prior values of y for the calculation. This approximation will be closer and closer to fact as T goes to zero; indeed, in the limit as T goes to zero the left-hand side of (7.37) will become a differentiation, and the whole equation will become (7.35).

If we take the z transform of (7.37) we get

$$\frac{1}{T}\left(Y(z)-\frac{Y(z)}{z}\right)\simeq a\left(U(z)-\frac{Y(z)}{z}\right); \tag{7.38}$$

this can be simplified to

$$\frac{z-1}{Tz}Y(z)\simeq aU(z)-a\frac{Y(z)}{z}. \tag{7.39}$$

It would seem, then, that we ought to be able to take the limit of (7.39) as T goes to zero and get a useful and meaningful expression that works as a transform for continuous-time signals. The difficulty in doing so is that we must find an expression for z that depends on T in such a way that it meets this goal.

To find a hint, let us start by solving (7.39) for our signal y:

$$Y\simeq\frac{aTz}{z-(1-aT)}U=\frac{(1-d)z}{z-d}U. \tag{7.40}$$

When u is a unit step we get

$$y(Tk)\simeq 1-d^k. \tag{7.41}$$

From this we get

$$y(t)\simeq 1-d^{\frac{t}{T}}=1-(1-aT)^{\frac{t}{T}}. \tag{7.42}$$

This is very suggestive: the approximate system has a pole at $z = d = 1 - aT$, and the system response has a component at d raised to t/T. In fact, in the limit as T goes to zero (7.42) goes to (7.36), where the system response involves e raised to $-at$. If we introduce a new transformed variable s and let

$$z=e^{-sT} \tag{7.43}$$

then we should have a reasonable approximation. Substituting (7.43) into (7.39) yields

$$\frac{e^{-sT}-1}{Te^{-sT}}Y(s)\simeq aU(s)-a\frac{Y(s)}{e^{-sT}}. \tag{7.44}$$

Equation (7.44) is just the *z* transform of (7.37); this means that (7.37) should become exact in the limit as *T* goes to 0:

$$\lim_{T\to 0}\frac{e^{-sT}-1}{Te^{-sT}}Y(s)=\lim_{T\to 0}aU(s)-a\frac{Y(s)}{e^{-sT}}. \tag{7.45}$$

This can be solved easily using L'Hospital's rule on the left-hand side:

$$sY(s)=aU(s)-aY(s). \tag{7.46}$$

Now we can solve for $Y(s)$:

$$Y(s)=\frac{a}{s+a}U(s). \tag{7.47}$$

In fact, (7.46) is just the Laplace transform of (7.35). We can extend this special case to derive the Laplace transform from the z transform:

$$L\{x(t)\}=\lim_{T\to 0}T\sum_{k=0}^{\infty}x(Tk)e^{-sTk}=\int_{0}^{\infty}x(t)e^{-st}dt. \tag{7.48}$$

Laplace and Differential Equations

Just as the z transform is used for solving shift invariant linear difference equations, the Laplace transform is used for solving time invariant linear differential equations. Like the z transform, the Laplace transform has certain useful properties and a family of "standard" transform pairs that can be used to solve many linear differential equations. Since continuous-time systems (such as plants) can often be modeled as linear time invariant systems, the Laplace transform is very useful.

Table 7.2 shows some common properties of the Laplace transform that can be used to solve differential equations. This table is similar to Table 2.1 in Chapter 2. Table 7.3 gives some common Laplace transform pairs, similar to Table 2.2 in Chapter 2. As with their counterparts for the z transform, these transform pairs and properties can be used to solve problems involving differential equations.

Property	Comments	
$L\{kx(t)\} = kX(s)$	*k* must be a constant.	(7.49)
$L\{x_1(t)+x_2(t)\} = X_1(s)+X_2(s)$	Together these imply that the Laplace transform is a linear operation.	(7.50)
$L\left\{\frac{d}{dt}x(t)\right\} = sX(s)$	As the unit delay is to the z transform, differentiation is the fundamental Laplace operation.	(7.51)
$L\left\{\frac{d^n}{dt^n}x(t)\right\} = s^nX(s)$	This is just the result of applying (7.51) multiple times.	(7.52)
$\lim_{t\to\infty} x(t) = \lim_{s\to 0}\left[\frac{1}{s}L\{x(t)\}\right]$	This is called the "Final Value Theorem" and is only valid if $x(t)$ settles to a finite value.	(7.53)

Table 7.2 Laplace transform properties

x(t)	X(s)	Comments	
$u(t)$	$\frac{1}{s}$	$u(t) = \begin{cases} 0 & t<0 \\ 1 & t \geq 0 \end{cases}$ [9]	(7.54)
$tu(t)$	$\frac{1}{s^2}$	$tu(t)$ is a unit ramp function.	(7.55)
$t^n u(t)$	$\frac{1}{n!s^{n+1}}$		(7.56)
$e^{-at}u(t)$	$\frac{1}{s+a}$	a must be a constant.	(7.57)
$te^{-atn}u(t)$	$\frac{1}{(s+a)^2}$	a must be a constant.	(7.58)
$\sin(\omega t)u(t)$	$\frac{\omega}{s^2+\omega^2}$	ω must be a constant	(7.59)

Table 7.3 Some common Laplace transforms

---

9 $u(n)$ is known as the "unit step function." Note that it *is* defined at $n = 0$.

The procedure for solving a linear time invariant differential equation with Laplace transforms is very like the procedure for solving shift invariant difference equations with z transforms. One applies the rules in Table 7.2 as necessary to translate the differential equation into the Laplace domain, then one uses the result as appropriate.

For example, take the differential equation that results from modeling the behavior of a DC motor being driven by a controlled voltage:

$$\frac{d^2}{dt^2}\theta(t) = -\left(\frac{k_T^2}{R_A J_A}\right)\frac{d}{dt}\theta(t) + \left(\frac{k_T}{R_A J_A}\right)v(t) \tag{7.60}$$

where k_T is the motor's torque constant in Newton-meters per amp, R_A is the armature resistance in ohms, J_A is the armature moment of inertia in kg-m^2, θ is the shaft angle in radians and v is the applied voltage. Applying (7.51) and (7.52) we get

$$s^2\Theta(s) = -\left(\frac{k_T^2}{R_A J_A}\right)s\Theta(s) + \left(\frac{k_T}{R_A J_A}\right)V(s). \tag{7.61}$$

The concept of transfer functions works in the Laplace domain as well as it does in the z domain, so we can find the transfer function for the motor:

$$\frac{\Theta}{V} = \frac{\dfrac{k_T}{R_A J_A}}{s\left(s + \dfrac{k_T^2}{R_A J_A}\right)} \tag{7.62}$$

This transfer function can be used directly for performing system analysis and design, or it can be converted into a z domain transfer function for use in a discrete-time system.

A Frequency Domain Interpretation

As with the z transform, the Laplace transform has a frequency-domain interpretation. In a manner similar to the development in Chapter 2, you can show that if you have a transfer function $H(s)$ and a sinusoid input with radian frequency ω the output, after the transients have died out, will be a sinusoid with amplitude and phase given by $H(j\omega)$.

7.8 z Domain Models

While Laplace-domain models are interesting, and reflect the reality of the continuous-time systems that we deal with, to do a discrete-time control system design, we ultimately need to bring our plant model into the z domain. There are a number of ways to do this; this section details the most popular methods.

Approximate Models

Often a control system design does not require an exact model, or there simply isn't enough information to generate an exact model. In such cases one may as well use a simple approximation to generate a model for preliminary design, then use measurements to refine the model as necessary for the final design steps.

To see how we may do this, first consider some simple ways to perform numerical integration. Of course you know that finding the integral of a function (or signal) means finding the area under the curve. So a signal $x(t)$ and its integral will look like Figure 7.13.

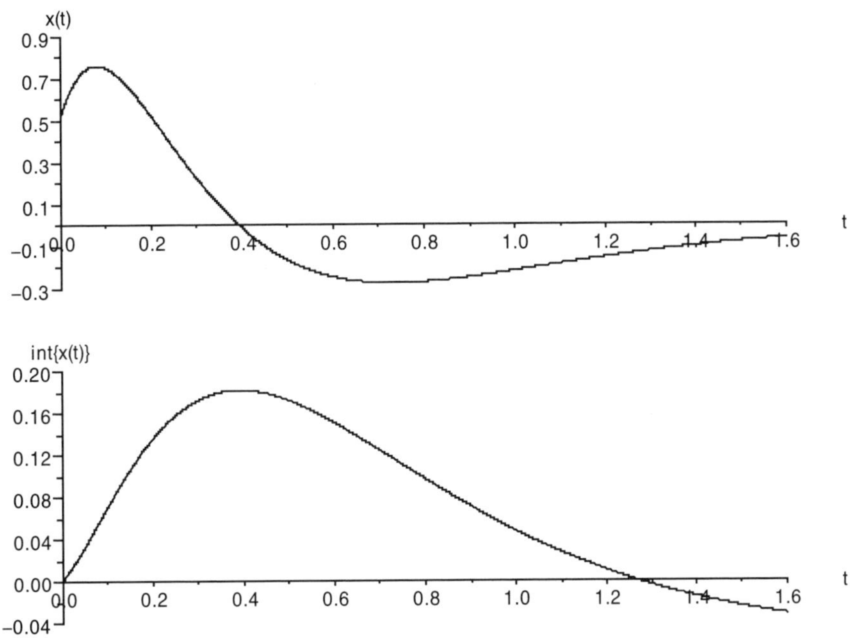

Figure 7.13 A signal and its integral

There are many methods of performing numerical integration; all of them involve chopping up the signal $x(t)$ in time, calculating an approximate value for each piece, and adding the areas together. There are a number of different algorithms, each one more clever than the last. The most obvious one is to use a constant time interval for each piece and to approximate the area of each piece as the time interval times the signal value at the beginning of the piece; this is illustrated in Figure 7.14, where $x(t)$ is integrated with a sampling interval of 0.1 second.

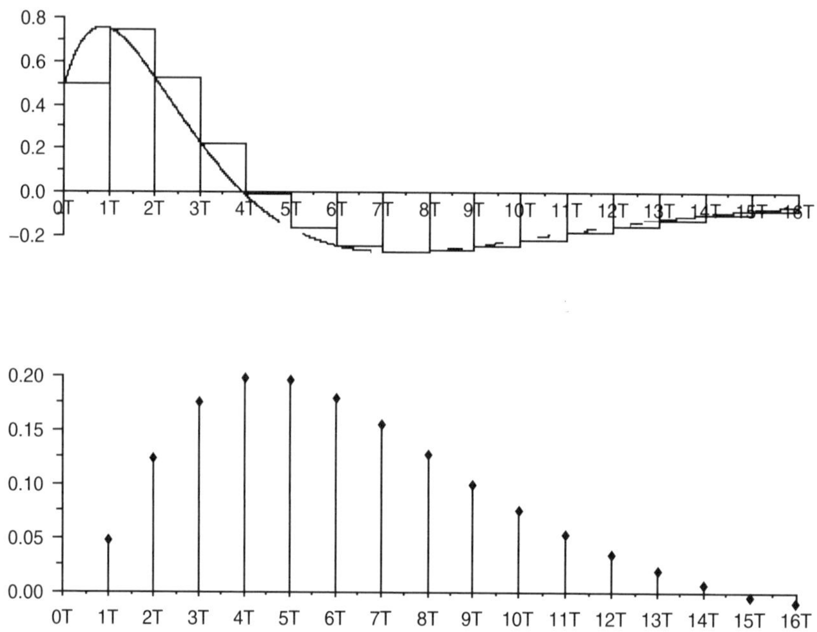

Figure 7.14 Integration using first forward difference

This technique is called the first forward difference for the way that one writes a difference equation from a differential equation when using it. The equation that expresses the integration in Figure 7.14 is

$$\int_{0}^{kT} x(t)\,dt \simeq T\sum_{q=0}^{k-1} x(kT) \tag{7.63}$$

(taking x as being zero for any $t < 0$). If you take the Laplace transform of the left side of this, and the z transform of the right,[10] you get

$$\frac{1}{s}X(s) \simeq \frac{T}{z-1}X(z). \tag{7.64}$$

If you take $X(s)$ as being approximately equal to $X(z)$ this can be further reduced to

$$s \simeq \frac{z-1}{T}. \tag{7.65}$$

You use the first forward difference approximation by replacing every occurrence of s in a Laplace domain transfer function with the right side of (7.65). So, for example, if you had a motor with transfer function

$$H(s) = \frac{\omega_0 / k_m}{s(s+\omega_0)} \tag{7.66}$$

you would replace this with

$$H(z) \simeq \frac{\omega_0 / k_m}{\frac{z-1}{T}\left(\frac{z-1}{T}+\omega_0\right)} = \frac{(\omega_0 / k_m)T^2}{(z-1)\left(z-(1-T\omega_0)\right)}. \tag{7.67}$$

This approximation has a severe disadvantage: while any unstable transfer function in the Laplace domain will correctly lead to an unstable approximation in the z domain, a *stable* transfer function in the Laplace domain can also lead to an unstable z domain approximation. For each pole of the Laplace-domain transfer function at $s = a$, there will be a pole of the z domain transfer function at $z = 1 - Ta$; clearly there are values of a that will cause this pole to be unstable even if the original transfer function was a stable one.

---

10 This is not as peculiar as it seems, but justifying why would go well beyond the scope of this book—just take it as an acceptable thing to do.

After rectangular integration (the type discussed above), the next most elementary approximation is trapezoidal integration. With trapezoidal integration the area of each slice of the integral is taken to be a trapezoid, as shown in Figure 7.15.

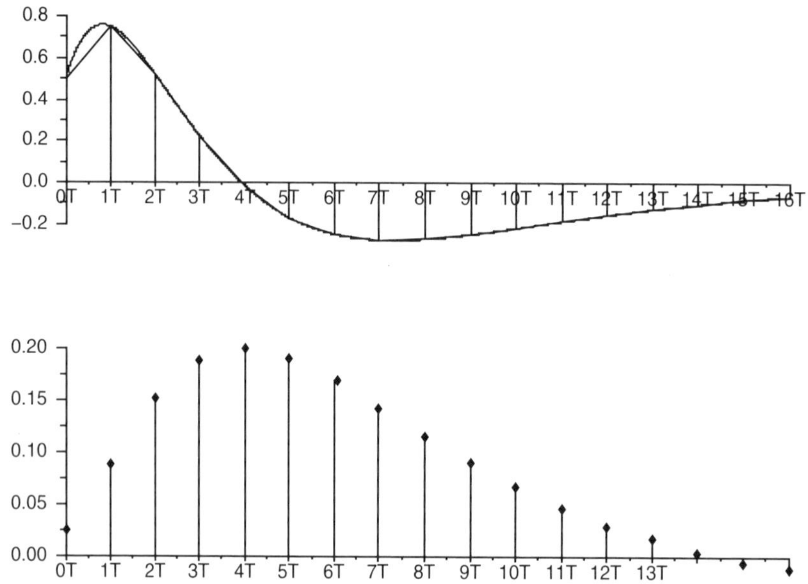

Figure 7.15 Trapezoidal integration

When trapezoidal integration is used the approximation becomes

$$\int_0^{kT} x(t)\,dt \simeq \frac{T}{2}\left[\sum_{q=0}^{k-1} x(qT) + \sum_{q=0}^{k} x(qT)\right] \tag{7.68}$$

keeping in mind the fact that x is zero for $t < 0$ we get

$$\int_0^{kT} x(t)\,dt \simeq \frac{T}{2}\sum_{q=0}^{k}\left[x(qT) + x((q+1)T)\right]. \tag{7.69}$$

Transforming both sides yields

$$\frac{1}{s}X(s) \simeq \frac{T}{2}\frac{z+1}{z-1}T(z) \tag{7.70}$$

or

$$s \simeq \frac{2}{T}\frac{z-1}{z+1}. \tag{7.71}$$

This approximation is known as the Tustin approximation.

If you make (7.71) exact and solve for z you will find that

$$z = \frac{2+Ts}{2-Ts}. \tag{7.72}$$

This identity has a very nice property: any stable pole in the Laplace domain maps to a stable pole in the z domain and vice versa, and any *un*stable pole in the Laplace domain maps to an unstable pole in the z domain (and vice versa). These pole locations will not be exact equivalents, but the Tustin approximation will always generate a stable transfer function in the Laplace domain from a stable transfer function in the z domain.

Consider the motor transfer function we used before. With the Tustin approximation this transfer function becomes

$$H(z) \simeq \frac{\omega_0 / k_m}{\frac{2}{T}\frac{z-1}{z+1}\left(\frac{2}{T}\frac{z-1}{z+1}+\omega_0\right)} = \frac{T^2\omega_0}{2k_m(2+\omega_0 T)}\frac{(z+1)^2}{(z-1)\left(z-\frac{2-T\omega_0}{2+T\omega_0}\right)}. \tag{7.73}$$

The Tustin approximation is generally useful when you must model a plant whose response is known in general but which will require measurements to actually tune the controller, and for times when you are converting an inherited analog controller to digital.

Exact

The Tustin approximation is a good way to get a close idea of how a plant or other continuous-time system will behave from the point of view of a digital system. However, not only is it an approximation, but it completely ignores the effects of sampling, reconstruction and controller delay on the signal.

Ideal samplers and hold blocks are linear; if you assume that your plant and controller are linear, then your model remains linear when you add sampling and reconstruction to it. While your sample and hold blocks are *not* time invariant, they *are* shift invariant to the controller, so from the point of view of the controller you should be able to generate a z domain model of the plant that incorporates the controller, sample, and hold blocks.

Figure 7.16 shows the system block diagram that we'll assume for deriving an exact z-domain model of a continuous-time system. The controller puts out a control signal $u_c(k)$ which is converted into the continuous-time signal $u_c(t)$ by the zero-order hold block. The plant (represented by the transfer function $G(s)$) responds with the continuous-time signal $y(t)$ which is sampled into $y(k)$ by the sampler. The model will incorporate the effect of sampling and reconstruction, and will be called $G(z)$.

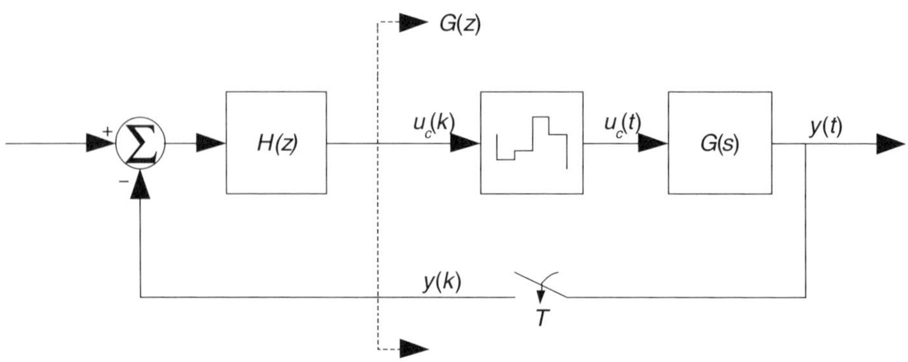

Figure 7.16 System for deriving exact model

Because the system is linear and shift invariant, we can find the transfer function of the plant if we can find its step response. If we apply a unit step to $u_c(k)$, then the output of the reconstruction filter, $u_c(t)$, is just a continuous-time step. So the modeling process becomes:

1. Calculate the step response of $G(s)$ in the Laplace domain (by multiplying $G(s)$ by $1/s$).
2. Transform this into the time domain.
3. Write the sampled-time version of this step response.
4. Transform into the z domain.
5. Find $G(z)$ by dividing the step response by the z-domain step $z/(z-1)$.

Example 7.3

A control system similar to the one in Figure 7.16 consists of a controller, plant, zero-order hold and a sampler. The plant is a heater and thermal mass whose transfer function has been simplified to

$$\frac{Y(s)}{U_c(s)} = \frac{k_h}{\left(\frac{s}{\tau_1}+1\right)\left(\frac{s}{\tau_2}+1\right)}. \tag{7.74}$$

The time constants are 10 and 50 milliseconds; the sampling frequency T is 200Hz. Find the transfer function of the plant as seen by the controller.

In the Laplace domain, the step response of the system is

$$Y_{step} = \frac{k_h}{s\left(\frac{s}{\tau_1}+1\right)\left(\frac{s}{\tau_2}+1\right)}. \tag{7.75}$$

This expands to

$$Y_{step} = \frac{k_h}{s} + \frac{\tau_2}{\tau_1-\tau_2}\frac{1}{s+\tau_1} - \frac{\tau_1}{\tau_1-\tau_2}\frac{1}{s+\tau_2}. \tag{7.76}$$

In the time domain this becomes

$$y_{step}(t) = \begin{cases} 0 & t<0 \\ k_h + \frac{k_h\tau_2}{\tau_1-\tau_2}e^{-\frac{t}{\tau_1}} - \frac{k_h\tau_1}{\tau_1-\tau_2}e^{-\frac{t}{\tau_2}} & t \geq 0 \end{cases}. \tag{7.77}$$

Converting this to the z domain yields

$$Y_{step}(z) = \frac{k_h z}{z-1} + \frac{k_h\tau_2}{\tau_1-\tau_2}\frac{z}{z-e^{-\frac{T}{\tau_1}k}} - \frac{k_h\tau_1}{\tau_1-\tau_2}\frac{z}{z-e^{-\frac{T}{\tau_2}k}}. \tag{7.78}$$

If we let

$$d_1 = e^{-\frac{T}{\tau_1}}$$
$$d_2 = e^{-\frac{T}{\tau_2}} \tag{7.79}$$

then

$$Y_{step}(z) = \frac{k_h}{\tau_1 - \tau_2}\frac{z}{z-1}\frac{\tau_1(1-d_1)(z-d_2) - \tau_2(1-d_2)(z-d_1)}{(z-d_1)(z-d_2)}. \tag{7.80}$$

Removing the step, we get just the transfer function

$$\frac{Y(z)}{U_c(z)} = \frac{k_h}{\tau_1 - \tau_2}\frac{\tau_1(1-d_1)(z-d_2) - \tau_2(1-d_2)(z-d_1)}{(z-d_1)(z-d_2)}. \tag{7.81}$$

Substituting in values for the time constants and sampling time we get

$$d_1 = e^{-0.5} \simeq 0.607$$
$$d_2 = e^{-0.1} \simeq 0.905$$
$$G(z) = k_h\left(\frac{0.021z - 0.017}{(z-0.607)(z-0.905)}\right). \tag{7.82}$$

7.9 Conclusion

This chapter covered sampling theory and the Laplace transform. It brings the material in this book into the real, continuous-time world, mostly by giving us methods for bridging our discrete-time control system descriptions with continuous-time plants.

8

Nonlinear Systems

So far, I have covered various ways to analyze linear systems using the z transform and sometimes the Laplace transform. I have not treated issues of nonlinear control in any systematic way. This is partially because nonlinear control is a difficult and slippery topic, and partially because up to this point I have been writing about control using the z transform, which is an inherently linear transform that is only useful for analyzing strictly linear systems.

The z transform, and other forms of linear analysis, can only go so far in designing control systems, for one simple reason: no real world system is linear. Everything in the world, when subjected to a strong enough input, will bend or break or burn or yield or stiffen up or otherwise exhibit nonlinear characteristics. This means that, strictly speaking, there are no real-world systems that satisfy the requirements of a linear system that can be analyzed using the z (or Laplace) transform.

Why, then, have we spent so much time and effort on these linear systems analysis methods? Because finding even partial and particular solutions to nonlinear differential equations is a thankless, sometimes difficult and often impossible task. Because the analysis of linear systems is relatively easy, we want to find ways that we can approximate our nonlinear systems as linear ones. Where we can't do that, we want to at least draw inspiration from solutions to linear problems when we must find controllers to nonlinear systems.

The general procedure for analyzing and designing control systems for nonlinear plants is to carry a large bag of tricks, and be ready to pull as many tricks out of the bag, as often as necessary, to find an adequate solution. You try to keep the tricks that work, discard the tricks that don't, and refrain from fooling yourself into thinking that you've attained higher performance than is really possible.

It is important to remember while you learn these tricks that "optimal" solutions to controlling nonlinear plants are few and far between. Those few times when you are lucky enough to encounter a nonlinear plant and performance requirement that allow you to develop an optimal solution, you will find that it is specific to that one circumstance, and will be useless on any problem that is not very similar. It is best to remember this when designing controllers for nonlinear systems, and be sensitive to the point when you have achieved enough performance to meet requirements. Continuing to work on the system design after you have made a high-quality robust system of sufficient performance may optimize for other things, but it certainly won't optimize for engineering time!

8.1 Characteristics of Nonlinear Systems

To apply knowledge about linear systems to nonlinear systems, one needs to understand the differences between them. This knowledge will then act as a guide to knowing where linear system theory can be used as-is, and where we must use nonlinear system knowledge to modify our conclusions.

A linear system is defined by the fact that the property of superposition holds, and many of the characteristics of linear system analysis flow from this one property. One of the important results of superposition is that the behavior of a linear system is global: it will always have the same relative response to a particular initial condition or input signal, no matter what the magnitude of the initial condition or input signal. This means that the complexity of a linear system's response is determined entirely by its order—for a system of order n, once you know its input signal and its first n outputs, you know everything you need to know to find its output for all time.

The obvious conclusion of this global property of linear systems is that nonlinear systems must not exhibit global behavior, and in fact this is true in general. This seems like a simple difference, even a quibble, but it is responsible for many, if not most, of the difficulties one encounters trying to control nonlinear systems.

Take the example of a simple pendulum as shown in Figure 8.1.

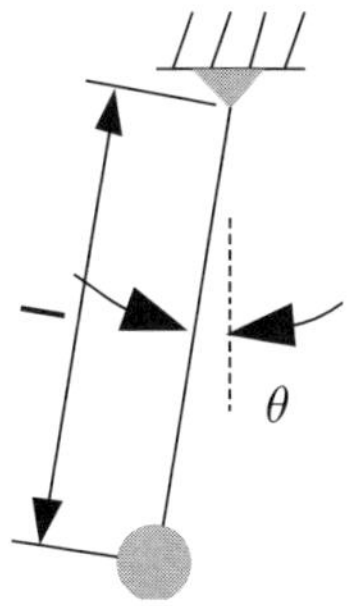

Figure 8.1 A simple pendulum

The equations of motion for this system are simple:

$$\frac{d^2}{dt^2}\theta = -\frac{g}{l}\sin\theta, \tag{8.1}$$

where g is the acceleration of gravity. Now, (8.1) is clearly nonlinear, but you can approximate it if you declare that θ will never vary too much from zero, then use the approximation $\sin\theta \simeq \theta$ to get

$$\frac{d^2}{dt^2} = -\frac{g}{l}\theta. \tag{8.2}$$

If you solve this equation you'll get the well known solution for harmonic motion observed by Galileo so long ago:

$$\theta(t) = A\cos\left(\sqrt{\frac{g}{l}}t + \phi\right). \tag{8.3}$$

This equation shows a sinusoidal motion with a constant period. If you were to plot the angle against the angular velocity, the result would be a perfect circle. For different starting points the circles would have different diameters, but otherwise they would all be identical.

The situation is different if you use the model in (8.1). Figure 8.2 shows the position versus velocity plots for a real pendulum that obeys (8.1). Notice that close to the (0, 0) point the trajectories are, indeed, circular. There are two discrepancies between this figure and what you would expect of a linear system, however: first, as you get away from the center point the trajectories start to distort, until finally the trajectories no longer follow closed paths at all, and second, there is more than one point around which circular trajectories are centered.

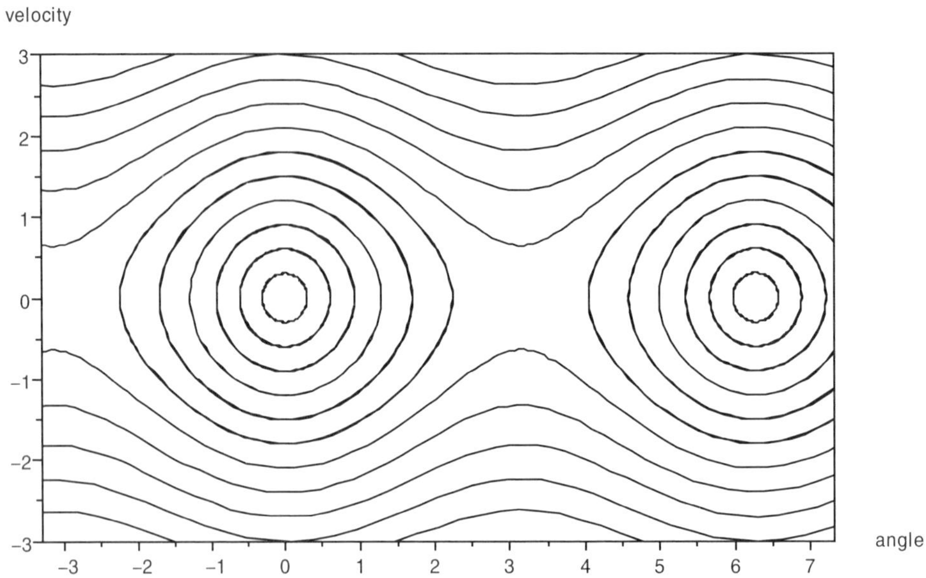

Figure 8.2 Phase plane plot of pendulum motion

The multiple trajectories in Figure 8.2 are simply a consequence of the way angles are expressed: there are circles centered on 2π, but for a real pendulum, that's the same as an angle of zero. What's more important is the trajectories on the top and bottom of the figure. At these extremes, the pendulum is no longer swinging. Instead, it has started spinning around the pivot, which is a quite different behavior indeed than swinging back and forth.

This example shows one of the problems faced by control systems designers in a nonlinear world: while the behavior of the pendulum close to (0, 0) is predictable and well behaved, moving the system off that point will result in significantly different behavior which must be accounted for if the system is to work in those conditions. The problems created by the nonlinear world range from insignificant (and often not noticed in practice) to insurmountable.

8.2 Some Nonlinearities

It would be futile to try to detail and classify all of the different types of nonlinearities one may encounter in the real world. Instead, I am just going to give some illustrative examples, in some overlapping types, make some comments about the properties of each one, and point to some of the ways that they may be dealt with.

Sneaky Nonlinearities

The worst sort of nonlinearity is the kind that you discover too late in the design process to deal with in a graceful manner. Nonlinearities that are not properly dealt with become bugs just like any other bug, and like any other bug, the earlier you know about them the less expensive it will be to deal with them. The consequences of a nonlinearity can range from the need for a minor retuning, to the discovery that an entire technical approach is invalid; obviously in the latter case, it's best to figure this out early in the process.

Of course, unconsidered nonlinearities are usually not big and obvious (although I have certainly been guilty of overlooking some that are, in hindsight, pretty big). Nonlinearities usually go unnoticed either because they only occur when a system is pushed to some extreme of its operating envelope, or because they are small but not so small that they can be disregarded when you're trying to squeeze that last essential bit of performance out of your system.

For an extreme example, consider a length of 12-gauge copper wire, suspended in free air. This most linear of components has a simple relationship between its current flow and the voltage along its length. Now apply 500 amperes to the wire for a significant fraction of a second. The wire will heat, which will cause its resistance to go up, increasing the amount of power it receives. It will soon melt or possibly vaporize, after which it will never again have the same current/voltage relationship. One doesn't normally consider melting wires to be a nonlinear effect—but they are. One doesn't normally consider exploding wires to be sneaky, but if you hadn't anticipated it they can certainly come as a surprise.

A less extreme example of unanticipated nonlinearities would be a mechanical system that has hard stops or snubbers at its ends of travel. Assume that you have a nice control rule that works perfectly well if the mechanism never reaches the end of its travel—can you be sure that this control rule will work adequately if the mechanism overshoots and bumps into the stop? What may happen if the target position is set beyond the stop? While the range of bad things that could happen will vary widely with the nature of the mechanism and the control rule, this is a use case that should be considered and the system behavior should be predicted, tested for, or the situation avoided.

For an example of a nonlinearity causing problems at the small, subtle end of things, consider a linear power amplifier. Such amplifiers are usually implemented as two amplifiers: one amplifier pushes current into the load, and another amplifier pulls current out of the load. Somewhere in the middle of its operating range, when it is switching from supplying power in one direction to supplying it in the other, a significant nonlinearity can happen. Called "crossover distortion" in audio circuits, it often takes the form of a dead spot in the amplifier's input/output relationship. Such a nonlinearity could cause severe performance degradation or even oscillation when the system is all settled out and stable—which is usually just where you want it to be on its best behavior.

These are just three types of sneaky nonlinearities, but just about any nonlinearity can sneak up on you; knowing what is possible and being willing to check all the plausible cases is the best defense against them.

Nonlinearities with Memory

In the section above I described how a melting wire could be considered to be a nonlinearity. It is also a nonlinearity that is tightly coupled to more than one system state: as the wire heats up, its resistance changes. First, the resistance changes temporarily when the wire is not hot enough to be damaged, but there is some time lag in the resistance going up or down. Second, if the wire does suffer damage, its resistance change will be permanent, and the amount of change will range from the slight to the dramatic.

Trying to think about such behaviors as one monolithic model can have you gibbering and climbing the walls. Fortunately you don't have to, because you can always split the system's nonlinearities from the system's memory and model them separately. In fact, one mathematically useful representation for a generic nonlinear dynamic system is to use a nonlinear state-space representation

$$\frac{d}{dt}x = f(x,u,t). \tag{8.4}$$

This equation shows the system state as the vector x with as many members as necessary to describe the system dynamics fully. The function f is, by definition, memoryless, although it may be nonlinear and/or time-varying. With this representation, we are not mixing the differential equation and the nonlinearities—the system description is split into the state vector which carries all the system memory, and a perfectly memoryless function of the current system state and its inputs.

This does not mean that a nonlinearity that is closely associated with a memory element will always be easy to deal with—if the only evidence that you can get of a nonlinearity comes to your controller after it has been integrated or otherwise filtered, it can be difficult or impossible to deal with it adequately. This does not mean that you can't model it, however.

In general, the problems caused by nonlinearities with memory aren't so much the memory element itself as the fact that the actual nonlinearity is buried, or that it is feeding back into the system in such a way as to make its effects hard to predict or control.

Figure 8.3 shows an example of the sort of system that may be encountered in practice: a nonlinear low-pass filter. The only available output comes from the integrator, not directly from the nonlinearity. Because of the integration, it can be difficult to determine just what the output of the x^2 term may be. Because of the x^2 term in the nonlinearity, the filter will not decay at a constant exponential rate, and indeed could even go unstable for large enough values of $u - y$. If it were necessary to operate the system in a manner that would let the value of $u - y$ over a wide range, then this system could become difficult to control in a reasonable manner.

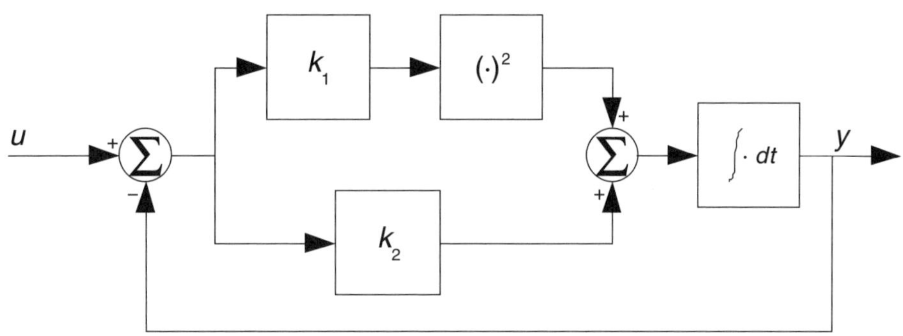

Figure 8.3 An example of buried nonlinearity

Surface Nonlinearities

The best kind of nonlinearity from an ease-of-design standpoint is a nice predictable nonlinearity that occurs right at an input or output and that doesn't affect the system dynamics in any significant way. Actuator or driver nonlinearities and output or sensor nonlinearities can be extremely benign, to the point where you can sometimes deal with them early in the design process and subsequently forget about them.

Actuator Saturation

Actuator saturation is the universal nonlinearity. If a control system is not asking everything from its actuators, it is because there is some intentional feature of the controller that is causing the limiting to occur somewhere else. Because actuator saturation is ubiquitous, control systems designers often apply measures against actuator saturation without considering that they are compensating for a nonlinearity. This means that they are missing the fact that they are applying solutions that may be useful in other situations, and that they are dealing with a problem that may have solutions they have not anticipated.

Any actuator that you may choose to drive your system, be it a motor, a heater, a solenoid or anything else, has some finite limit to the amount of drive it can take before overloading somehow. To deal with this, you choose drive electronics that limit the strength of the drive to the actuator in order to prevent excess smoke and metal pieces in your product.

In systems with actuator saturation the output must follow

$$u_{actual} = \begin{cases} u_{max} & u_{desired} > u_{max} \\ u_{desired} & u_{min} \le u_{desired} \le u_{max} \\ u_{min} & u_{desired} < u_{min} \end{cases}. \tag{8.5}$$

If you have an otherwise linear system with actuator saturation and you model it as a purely linear system, the model will be exactly accurate until the moment that the drive hits its limit. At this point your linear model will become drastically inaccurate. In such a case, the results of hitting this limit can range from the perfectly innocuous to the disastrous.

A detailed discussion of compensating for actuator saturation is given in "Restrict the System State," following.

Friction

Friction, even though its results are often fairly easy to observe, can be a pernicious and difficult nonlinearity to deal with—to the point where mechanical systems that must achieve high accuracy over large distances must often be designed with two locating mechanisms per axis—one being a long-stroke "coarse" mechanism with fairly high friction and the other a "fine" mechanism, specially built to eliminate friction as much as possible and having just enough stroke to cover the uncertainty of the coarse axis.

Figure 8.4 shows a typical plot of the force generated in a sliding joint as its velocity varies. Friction tends to have three components: sticktion, where the joint requires some amount of force to break away and start moving; Coulombic (dry) friction, which is an essentially constant force opposing the motion, and viscous friction, which is proportional to the speed between the two sliding members.

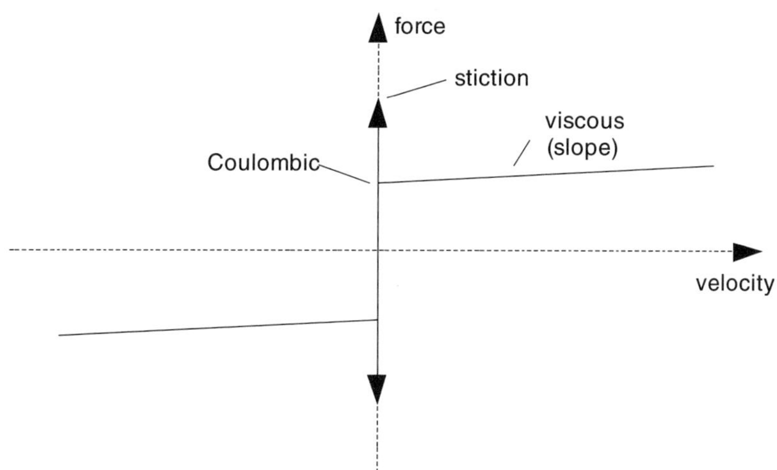

Figure 8.4 Typical plot of friction vs. velocity

Plants with significant amounts of friction can present some very difficult control problems, particularly if one naively assumes that a simple linear controller will do the trick. The joint has essentially zero gain for any force less than the threshold; this transitions to an effective negative gain when the stiction lets loose and the joint starts moving. The result is that without some specific friction compensating techniques, a PID controller driving a plant with friction will invariably enter a limit cycle where the integrator will slowly build up until the stiction is overcome, the mechanism will jump, then come to a stop (never at the target), stiction will reestablish itself, and the cycle will repeat.

There are two popular methods for overcoming the problems due to friction: pulsating drive (commonly called PWM) and deadband on the error signal.

Hysteresis

Hysteresis is a special form of nonlinearity with memory. A system with hysteresis will have an input/output relationship similar to Figure 8.5. The output signal will track the input signal whenever the input signal has been moving in one direction for a while, but should the input signal reverse direction there will be a range of input values where the input doesn't affect the output. If the input continues traveling in this new direction it will once again couple to the output and the two signals will once again track each other.

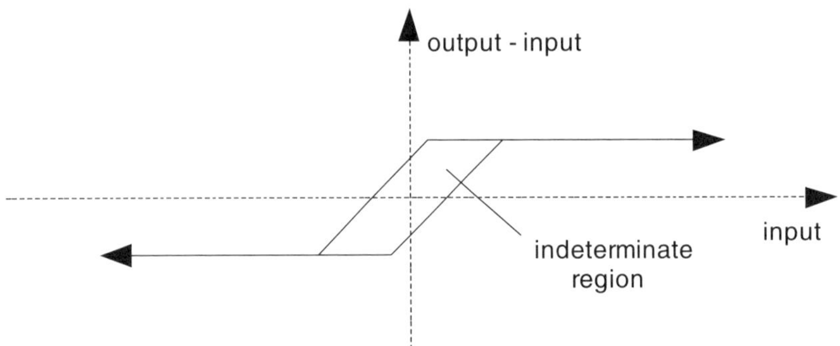

Figure 8.5 Hysteresis

A classic example of hysteresis is backlash. Backlash is a phenomenon that is usually seen when one is using gear trains; however, it can also occur when the output of a plant is driven through linkages. Backlash happens when there is some free play in a coupling between an input and output such that when the mechanism is in its backlash zone, there is no effective coupling between the input and output. When this happens, parts of the system become effectively invisible to the controller, usually to the detriment of stability, or at least to the controller's capability to control the system accurately.

Backlash is not the only effect that can cause hysteresis. A system that has some friction that is driven by a springy or soft mechanism will have hysteresis. Iron core transformers have a form of electromagnetic hysteresis in the core. Piezoelectric actuators will often show some hysteresis.

Because of the indeterminacy of the position of the input versus output, a system with hysteresis can be difficult to drive accurately. One is often constrained in such cases to reduce one's expectations to what the system is capable of achieving. Trying to force more performance out of such a system can lead to systems that always tantalize but never continue to work correctly for any length of time.

Hysteresis is generally dealt with by adding some deadband to your controller.

8.3 Linear Approximation

The most common method of solving nonlinear control problems is to pretend that you're solving a linear control problem. The general method to do this comes in five steps, only the first two of which many designers perform (or are aware). These steps are: first, find a linear system model of the nonlinear system that appears to be good enough for analysis and design; second, design your controller as if your system really were linear; third, embellish your linear model with features that make it more accurate yet still easy to analyze; fourth, embellish your controller design with features to keep the system design within the stated parameters; and finally go back and make sure that you really hit the mark.

Each of the four methods outlined in this section is a different way to take a nonlinear system and pretend that it is a linear one. Used with care they can allow you to use the tools for system analysis and design that you already have to work with many nonlinear systems.

Linearize around the Operating Point

Conceptually, the easiest way to find a linear approximation for a nonlinear system model is to choose an operating point and differentiate the model around that point. This method of solving nonlinear problems is pervasive; many people use it without realizing that they are doing so, and consequently they turn all of the nonlinearities in their system into sneaky nonlinearities.

Linearizing around an operating point doesn't even have to be applied to the system as a whole: the bits and pieces of the model can be linearized individually as I did when I went from (8.1) to (8.2). When you use this piecemeal approach to linearization, however, you must be careful to remember what you have done, and to take into account that you are piling approximation on top of approximation.

The general method for performing linear approximation is to inspect the model once it is done, or to inspect the pieces as you assemble them into a model, and find a convenient way to approximate the model as a linear system.

Take the temperature control system

$$T_{process}(t) = h_{temperature}\left(v_{drive}^2\right) \tag{8.6}$$

where $T_{process}$ is the plant temperature, v_{drive} is the voltage to the heater, and $h_{temperature}$ is a linear system that represents "the rest" of the plant. Now assume that you know through experiment or calculation that the normal voltage that will be required is v_{drive} = 16V. Then you can approximate the v^2 term:

$$v_{drive}^2 \simeq v_o^2 + v_{drive} \frac{d}{dt} v_{drive}\bigg|_{v_{drive}=v_o} \tag{8.7}$$

or

$$v_{drive}^2 \simeq (16V)^2 + v_{drive}\left((2v_{drive}\,_{v_{drive}}=16V\right) = 256V^2 + (32V)\,v_{drive} \tag{8.8}$$

and the system equation reduces to

$$T_{process}(t) \simeq h_{temperature}\left(T_{ambient}, (32V)\,v + 256V^2\right). \tag{8.9}$$

As long as the drive voltage settles to something close to 16V, then (8.8) and (8.9) will be valid.

As another example, consider a speed controller for a universal-wound motor. These motors have a torque that is proportional to the square of their terminal current, which is in turn a function of the terminal voltage and motor speed:

$$T_a = ki_t^2, i_t = \frac{v_t}{R_t + \omega_a k} \Rightarrow T_a = k\left(\frac{v_t}{R_t + \omega_a k}\right)^2 \tag{8.10}$$

where T_a is the armature torque, k is the motor winding constant, i_t is the terminal current, v_t is the terminal voltage, R_t is the terminal resistance and ω_a is the armature speed. The armature speed, in turn, depends on torque:

$$\frac{d}{dt}\omega_a = \frac{T_a(v_t, \omega_a)}{J_a} = \frac{k}{J_a}\left(\frac{v_t}{R_t + \omega_a k}\right)^2. \tag{8.11}$$

Here again we have a differential equation that is quite nonlinear, not just in the input (because v_t is squared), but in the states as well, because the acceleration depends on a nonlinear function of speed. This model can be linearized around nearly any operating point, however, just as we did for (8.6):

$$\frac{d}{dt}\omega_a \simeq \left.\frac{T_a(v_t, \omega_a)}{J_a}\right|_{\substack{v_t = v_{0,} \\ \omega_a = \omega_0}} + \left.\frac{d}{dv_t}\frac{T_a(v_t, \omega_a)}{J_a}\right|_{\substack{v_t = v_{0,} \\ \omega_a = \omega_0}} + \left.\frac{d}{d\omega_a}\frac{T_a(v_t, \omega_a)}{J_a}\right|_{\substack{v_t = v_{0,} \\ \omega_a = \omega_0}}, \tag{8.12}$$

which resolves to

$$\frac{d}{dt}\omega_a \simeq \frac{k}{J_a}\left(\frac{v_0}{R_a + \omega_0 k}\right)^2 + \frac{2kv_0}{J_a(R_a + \omega_0 k)^2}v_t - \frac{2k^2 v_0^2}{J_a(R_a + \omega_0 k)^3}\omega_a. \tag{8.13}$$

This equation is complex, but is at least linear.

Describing Functions

Some situations are not a good fit for linearizing a system around an operating point. A system may have sharp nonlinearities, the anticipated swing in the input may be too great for the approximation to be accurate, or the plant characteristics may simply not be well known. This doesn't mean that the system cannot be described as a linear system for which you can develop a controller using linear techniques. The challenge becomes finding a way to get the linear system. One good technique—and one that lends itself to using real-world measurements in your system design—is describing function analysis.

In describing function analysis you either take a detailed nonlinear model of your system and simulate it while injecting some signal that's of interest to you, or you take the *real* system and inject your signal of interest while looking at the system response. Usually, the signal you inject is a sine wave or a step, the former because it lends itself to frequency-domain analysis and the latter because it is very representative of the kind of input your system will get. You take the input and output signals from the simulation or measurement and you use the results to derive a linear system model using system identification techniques,[1] then you use that model to design your controller.

Describing function analysis has several features in its favor: it will work for any system that can be stabilized and measured, not just systems that are candidates for direct linearization techniques; it can be done from measured data rather than a mathematical model so it can be more accurate; and it will often reveal interesting system behaviors that are difficult to predict or understand otherwise.

Describing function analysis also has some pitfalls that must not be overlooked. First, the models derived using describing function analysis are only exact if the exact controller used in the actual system is used—to the point where you may need to redo your measurement or simulation whenever you make a significant change in your tuning. Second, the use of describing function analysis doesn't guarantee performance or even stability. Third, because the system is nonlinear,

[1] System identification involves extracting a system model from measured (or simulated) data. I won't treat it fully in this book, but the swept-sine measurement technique detailed in Chapter 9 is a form of system identification that is very useful in practice.

the simulation (or measurement) must often be performed with inputs at a number of different amplitudes—just one measurement would misrepresent the behavior of the system. Finally, if the nonlinearities are ones that can cause signal distortion, you will need to be aware that this distortion will appear in the output to the system, so you can't expect a sine wave output for a sine wave input.

Model with Uncertain Parameters

For both the case of using a continuous approximation or with describing function analysis, the apparent parameters of the linear system will change with the operating point and with the amplitude of the input signal. If you tried to model your system with an explicit dependence between the parameters and the operating point then you will be back to a nonlinear model to which you cannot apply linear design techniques.

What you can do, however, is to capture the amount of variation in your system model, either by calculation in the case of linearization or by multiple measurements in the case of describing function analysis. Then use this information about apparent variation to design a control system that will be robust to the changes.

An example of this method can be seen using the thermal control system described by (8.7): the apparent gain of the heater is directly proportional to the voltage on the heater, so a great change in heater voltage will cause a great change in the system dynamics. If you have a good idea of how much the heater's operating point will vary as the system is used you can design your control system to work over the whole range. To make sure that the heater will settle at the correct operating point, you can use describing function analysis or a number of simulations to verify correct operation.

Model with Apparent Noise

Some nonlinearities do not cause gross divergences of the system behavior from their linear models. In particular, nonlinearities such as the cogging from a geared drive, the cogging of a DC motor, or the discrete steps of an analog-to-digital (or digital-to-analog) converter will not upset the large-signal behavior of the system, but will cause small discrepancies in the system behavior that can impact the achievable precision of a high-performance system.

In such cases, the nonlinearity does not cause large instabilities; instead, the nonlinearity acts as if the system were otherwise linear but with a disturbance introduced into the system at the point of the nonlinearity. Figure 8.6 is an example of the input/output relationship you might see on a gear train with cogging. The large-scale relationship is linear, but there is a fixed-size component that diverges from the ideal linear relationship.

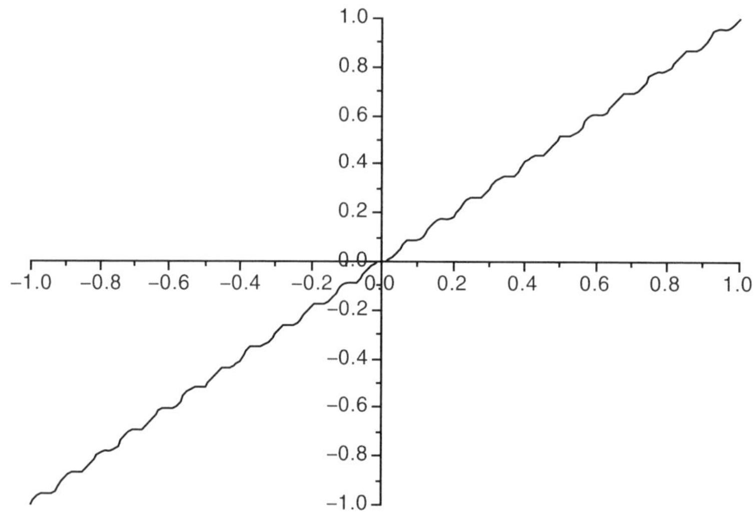

Figure 8.6 An example of cogging in a gear train

Nonlinearities of this type can be modeled fairly easily as a linear element plus a noise source. While the "noise" in this case is not random at all, treating it as such will simplify the analysis. As you do the analysis of your system, you have to make some choices about how you treat the noise: you must decide whether to assume that the noise appears as the RMS value of the discrepancy or if you should take the peak-peak value, and you must decide if you want to treat the noise as white (i.e., spread out over the frequency band) or as occurring at the worst possible frequency. It is most conservative (and often most accurate) to assume the worst-case, which is that the entire noise signal will appear at the worst possible frequency to wreak havoc on your system design. If your design will stand up to the effect at its worst, it will stand up to almost anything the real nonlinearity will dish out.

8.4 Nonlinear Compensators

This section is a compendium of the most useful tricks that I know for analyzing and/or designing controllers for nonlinear plants. These tricks can be loosely classified into tricks that let you pretend that you're controlling a linear plant, and tricks that put specific nonlinearities into your controller to compensate for specific nonlinearities in your plant.

None of these tricks are foolproof, and their use is much more "b'guess and b'gosh" than they are constructive. Even when one gets to the height of academic control theory, there aren't any widely accepted constructive methods of solving nonlinear control problems—just an ever larger bag of ever more elaborate tricks.

Inverse Functions

As stated earlier, the easiest kind of nonlinearity to deal with is a nice predictable nonlinearity that exists on the surface of the system (either at the input or output).

An almost trivial method to reduce or eliminate the effect of surface nonlinearities is possible when they embody an invertible function and when they remain predictable in the face of noise and all the usual variations. In this case you can (with appropriate cautions) simply apply the function's inverse to get an input or output function that is linear, at least over some range.

Formally, this is demonstrated in two parts. If you have a memoryless nonlinearity on a plant's output variable which can be expressed as

$$y = f(x), \tag{8.14}$$

then if f is an invertible function you can extract your desired input as

$$\hat{x} = f^{-1}(y). \tag{8.15}$$

Similarly if the nonlinearity is at the plant input then you would induce your desired drive by setting your drive to

$$x = f^{-1}(\hat{y}). \tag{8.16}$$

Of course, this affects your ability to accurately measure (or drive) your plant: if the function that you're inverting has a great variation in slope then errors will be greatly magnified where the slope of one or the other function is high. This method is also subject to problems when the inverse function that you have built into your controller doesn't match what's actually in the hardware—as with any other aspect of the plant, your nonlinear characteristics can vary; when both your approximation and the plant itself are sensitive to variations you can get some pretty bad answers pretty quickly if you're not careful.

Figure 8.7 gives an example of a function that one may encounter on the output of a plant—in this case it is

$$y = f(x) = x - \frac{\sin \pi x}{\pi}. \tag{8.17}$$

This function generates a one to one mapping between x and y, but its slope is zero at $x = 0$, which means that the slope of its inverse is infinite at $y = 0$. Using this function to read a value around $x = 0$ would be extremely noise sensitive compared to readings around the endpoints.

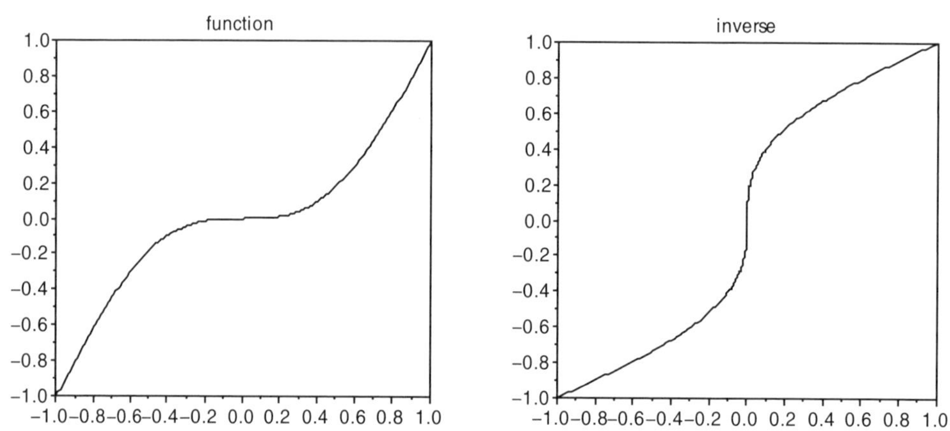

Figure 8.7 A function and its inverse

If you do contemplate using this method, you should do a little bit of analysis to make sure that it will fit with the precision and variability of the components that you planning on using. You should make sure that you won't have noise problems, or "glitches" in the corrected curve that are bad enough to cause you problems.

Gain Scheduling

It is not uncommon to have a nonlinearity that occurs in the form of two system states multiplied together, or a gain in a linearized model that varies with a system state. In either case, the net effect is that the system's linear model will be otherwise linear, but will have parameters whose values are dependent on some system state. When this happens, and when the governing state can be reliably measured or estimated, gain scheduling can be used to improve system performance.

Gain scheduling simply means that to compensate for the system's dependence on one of its states we change the parameters of our control system (and sometimes its structure) based on our knowledge of that state.

A classic example of gain scheduling is in the control of an aircraft. Aircraft flight surfaces generate moments that are roughly proportional to the square of the airspeed. For instance, the pitching moment due to the elevator angle on an aircraft would depend on the elevator angle and on airspeed:

$$T_{pitch} = kv^2\theta, \tag{8.18}$$

where k is a constant that depends on the airframe geometry, v is the airspeed, and θ is the elevator angle. If the craft's airspeed were to be kept constant then we could just linearize around that operating point, design a controller, and be done with it. In fact, some aircraft autopilots do just that—they are designed to only work through a specified range of airspeeds, and the pilot is expected to disengage them and take over control of the aircraft when flying outside of the safe flight envelope for that controller.

Aircraft, however, must be able to fly over a wide range of airspeeds. A fixed-parameter controller that would be superior for landing the craft may drive the system into instability at cruising speeds, and a controller that was adequate for cruising speeds may be hopelessly sluggish on landing. If the aircraft must fly with a controller on at all times this problem must be solved, and it is often solved with gain scheduling.

One way of arranging a gain-scheduling controller in this situation would be to command the elevator angle depending on the desired rate of rotation in pitch, i.e.,

$$\theta_{target} = \frac{\alpha_{pitch}}{kv^2_{measured}}. \quad (8.19)$$

This explicit method would work well, but would be quite involved to bring through an aircraft software certification process. Another way of implementing the gain scheduling would be to design as many sets of controller parameters as necessary to cover the speed range of the aircraft, and have the controller switch from one to the next as the speed changes. This method has the drawback of having a number of potential discontinuities in the system response, but it has the advantage that each individual set of parameters can be verified independently over its specified desired operating range.

Restrict the System State

Any of the above techniques will be effective for systems that are operated under certain conditions, with the inputs held to certain limits and when the system states are close enough to the nominal states under which the system was modeled.

At times one can design a linear controller using one of the above techniques and have its performance be quite stable and satisfactory over the entire operating range of the system; no matter what you do to the inputs of the system, the controller will be able to pull it back where it belongs in a smooth manner.

But often a linear controller that looks to be perfectly plausible when one considers a linear plant model can have severe problems if it strays from its operating point. Such a system can be unstable outright, it can be marginally stable, it can have objectionable sensitivity to noise—in some cases a system can even get itself stuck into a state that it cannot, on its own, get out of. If you are dealing with such a system and you know that the system will always work well if it stays in some restricted range of states then a good answer to this problem is to simply not allow the system to stray from the good range of states.

An example of restricting the system states can be seen in a fairly generic motion control system. Assume that a motion control system is to be designed around a linear "speaker coil" actuator that has an approximate discrete-time transfer function of

$$G(z) = \frac{1}{2T^2} \frac{z+1}{(z-1)^2}. \tag{8.20}$$

The control system will be sampled at 250Hz. In order to get satisfactory disturbance rejection a PID controller is to be used, with a transfer function of

$$H(z) = k_p + \frac{(1-d_d)(z-1)}{z-d} k_d + \frac{1}{z-1} k_i . \tag{8.21}$$

The goal of the control system is to close at about 10Hz, which gives controller constants of

$$\begin{aligned} d &= 0.25 \\ k_p &= 400 \\ k_i &= 1 \\ k_d &= 10000 \end{aligned} . \tag{8.22}$$

Both the plant's and the controller's transfer functions are normalized to ±1 times full scale. Both devices are limited to this amount; the controller because its drive electronics cannot support more current, and the plant because it runs into its stops.

A purely linear analysis of the system will, of course, always yield the same shape of response to a step input, no matter how big. Since we are dealing with a nonlinear problem, we must take these nonlinearities into account. Generally, this is done by simulating the system.

Figure 8.8 shows the motor/controller system simulated starting from an offset of –0.02 (1% of the total motion). Even with this small amount of travel the drive saturates at +1, although the system recovers nicely and settles the way you'd like it to.

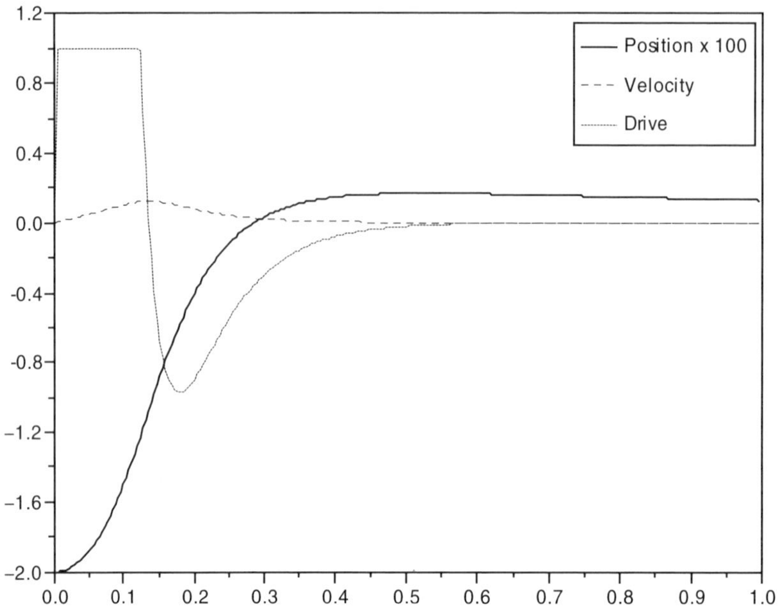

Figure 8.8 Motor/controller starting from position = –0.02

Figure 8.9 shows the exact same system, but this time the starting point is at –0.5 (25% of the total range). With this larger starting position, the system has gone from one that settles quickly (note the change in the time scales between the two figures) to one that goes into a large, slow oscillation. This type of oscillation can allow the equipment to reach high speeds and hit its end stops with potentially damaging force, or to apply continuous high drive to actuators that may not be able to dissipate the resulting heat. Even if nothing is getting broken or burnt up, it still isn't doing its job. Such behavior tends to alarm innocent bystanders and certainly isn't a good showcase for one's talents.

So why does the system behave this way? The reason is that the linear controller was designed assuming that there would be an infinite amount of drive available for slowing the system down. As a result, the controller just keeps pushing the system forward until it is almost where it's supposed to be, then generates an internal command that may be hundreds or even thousands of times larger than the maximum available drive.

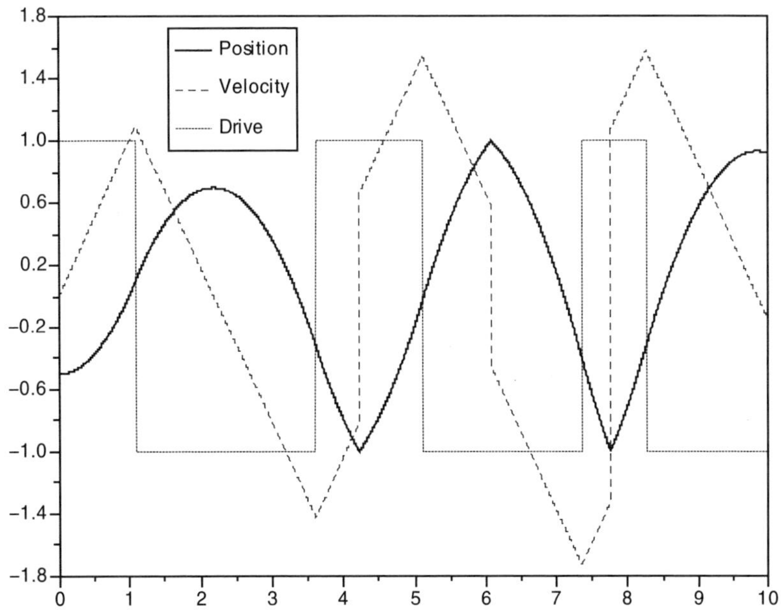

Figure 8.9 Motor/controller starting from position = −0.05

This result can be seen if you subject the system to describing function analysis. The linearized system has the Bode plot shown in Figure 8.10. Because the system has a triple integrator (two in the plant and one in the controller), it has enough low-frequency phase shift to give it both a high and a low gain margin. This means that if the system gain drops by too great an amount, the system will go unstable.

The essential nonlinearity in the system is the drive limit. If you were to feed a sinusoidal drive command into the system right before the limiter, then the output would be

$$u_{lim} = \begin{cases} 1 & A\cos(\omega t) > 1 \\ A\cos(\omega t) & -1 \le A\cos(\omega t) \le 1 \\ -1 & A\cos(\omega t) < -1 \end{cases}. \tag{8.23}$$

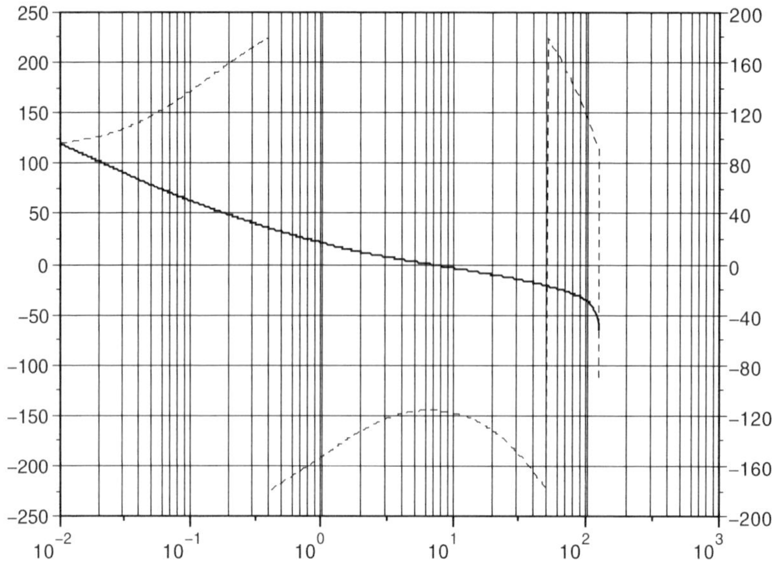

Figure 8.10 Bode plot of motor/controller system

Clearly, you can see that the resulting signal will have smaller magnitude than the input signal, and some thought will show that the phase of the signal is undisturbed. In fact, the magnitude of this signal's first harmonic is

$$A_{lim}(A) = \begin{cases} A & A \le 1 \\ 1 + \left(\dfrac{4A-2}{\pi}\right)\sqrt{A^2-1} - \dfrac{2}{A\pi}\cos^{-1}\left(\dfrac{1}{A}\right) & A > 1 \end{cases} . \tag{8.24}$$

For the purposes of describing function analysis, the effective gain of the limiter block is simply

$$H_{lim} = \frac{A_{lim}}{A} = \begin{cases} 1 & A \le 1 \\ \dfrac{1}{A} + \left(\dfrac{4A-2}{A\pi}\right)\sqrt{A^2-1} - \dfrac{2}{A^2\pi}\cos^{-1}\left(\dfrac{1}{A}\right) & A > 1 \end{cases} . \tag{8.25}$$

A little bit of analysis (or experimentation) will show that this gain is 1 until the signal magnitude exceeds 1, at which point it diminishes sharply with increasing signal magnitude. Figure 8.11 shows the gain profile as a function of the magnitude A.

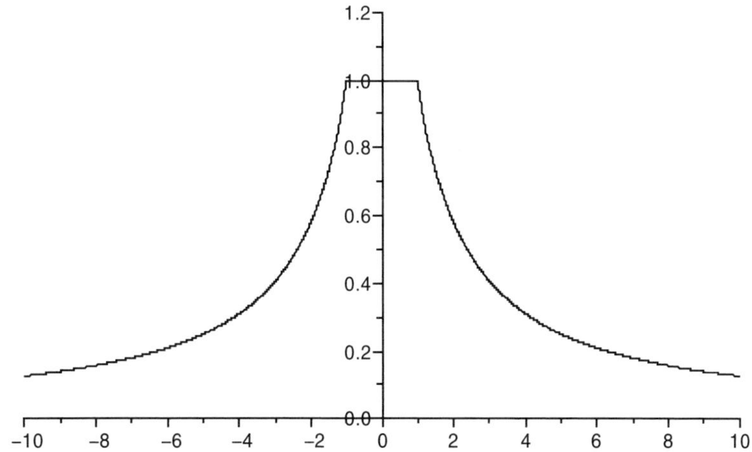

Figure 8.11 Effective gain of hard limited sine

Given this ever-diminishing gain with amplitude, it's easy to see why the system goes unstable. For high levels of required drive, the effective gain drops below the gain required for stability, and the system oscillates.

Command Profiling

So what can we do about this situation? The solution that is most commonly used in industry is to be careful about what we ask the system to do. We start solving this by observing that as long as the change in target position isn't too great, the system remains in its stable region. So if we find a way to issue the position command such that the drive doesn't bump into its maximum value (at least not for too very long), we should be OK. This approach effectively limits the system states to their "safe" values.

This is done by pre-calculating the velocity profile for the move, integrating it and issuing the resulting position command to the control system. The velocity profile is calculated based on a maximum allowable acceleration, a maximum allowable velocity, and the distance that one wishes to travel; a typical velocity profile looks like the one in Figure 8.12 and is called a *Trapezoidal* profile after its shape.

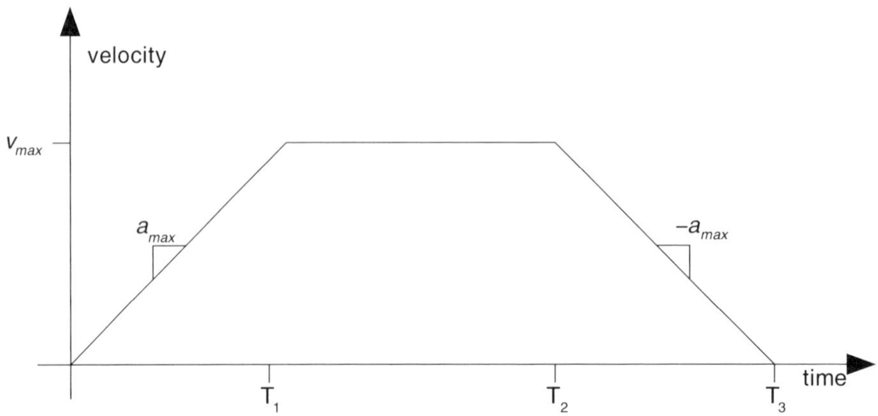

Figure 8.12 A trapezoidal velocity profile

The trick to designing a successful system using a trapezoidal velocity profile is to make sure that we will never demand more of the system than it can deliver. This means that we should ensure that the maximum requested accelerations and velocities are below what the system will be capable of over its entire operating range and as it ages. For a system that is going to be well maintained, it is safe to set the maximum allowable acceleration and velocity to 90% of what you expect the system can handle; for systems that need to be able to stand considerable wear and tear, you must reduce this figure according to your best judgment.

The points T_1, T_2, and T_3 are called the switching times. These switching times must be selected so that at the end of the planned trajectory the position command has traversed just the right distance. To do this, assign the distances d_1, d_2, and d_3 to the three switching times. Then we can find

$$d_1 = a_{max} \frac{T_1^2}{2} \tag{8.26}$$

from the distance traveled by a body under constant acceleration. The next distance is governed by the maximum velocity, so

$$d_2 - d_1 = (T_2 - T_1) v_{max} \, . \tag{8.27}$$

For the ending velocities to be the same, we must constrain the acceleration time and the deceleration time to be the same, i.e., $T_1 = T_3 - T_2$. One could do a whole bunch of calculations to determine that if this is so then

$$d_3 - d_2 = d_1 \, , \tag{8.28}$$

or one could just assert that it must be so by symmetry—I'll do the latter. When you add up these three incremental distances you get

$$d_3 = a_{max} T_1^2 + v_{max} (T_2 - T_1) \, . \tag{8.29}$$

In order to hit the maximum velocity you must use the relation

$$(T_3 - T_2) = T_1 = \frac{v_{max}}{a_{max}} \, . \tag{8.30}$$

Substituting (8.30) into (8.29) gives

$$d_3 = \frac{v_{max}^2}{a_{max}} + v_{max}\left(T_2 - T_1\right), \tag{8.31}$$

we can solve this to get

$$T_2 - T_1 = \frac{d_3}{v_{max}} - \frac{v_{max}}{a_{max}} \tag{8.32}$$

or

$$T_2 = \frac{d_3}{v_{max}} \tag{8.33}$$

(8.30) and (8.32) will give you the time points that you need to calculate your velocity profile unless (8.32) results in a negative answer. In that case you must use $T_2 = T_1$ and

$$\left(T_3 - T_1\right) = T_1 = \sqrt{\frac{d_3}{a_{max}}}. \tag{8.34}$$

Taken together this gives us

$$T_1 = \begin{cases} \sqrt{\dfrac{d_3}{a_{max}}} & d_3 < \dfrac{v_{max}^2}{a_{max}} \\ \dfrac{v_{max}}{a_{max}} & otherwise \end{cases}, \tag{8.35}$$

$$T_2 = \begin{cases} n/a & d_3 < \dfrac{v_{max}^2}{a_{max}} \\ \dfrac{d_3}{v_{max}} & otherwise \end{cases} \tag{8.36}$$

and

$$T_3 = \begin{cases} 2\sqrt{\dfrac{d_3}{a_{max}}} & d_3 < \dfrac{v_{max}^2}{a_{max}} \\ \dfrac{v_{max}}{a_{max}} + \dfrac{d_3}{v_{max}} & otherwise \end{cases}. \tag{8.37}$$

The resulting profile that we must generate is

$$d(t) = \begin{cases} 0 & t \le 0 \\ a_{max}\dfrac{t^2}{2} & 0 < t < T_1 \\ d_1 + v_{max}(t - T_1) & T_1 \le t \le T_2 \\ d_3 - a_{max}\dfrac{(T_3 - t)^2}{2} & T_2 < t \le T_3 \end{cases} ( ). \tag{8.38}$$

When this profile is used to generate position commands to the system, its behavior improves dramatically. Figure 8.13 shows the results of simulating the exact same system that generated Figure 8.9, but using a position command generated from (8.38).The resulting drive command does hit its limits, but the motor is always able to catch up in a smooth manner, and the system as a whole behaves in much the way that you would like it to.

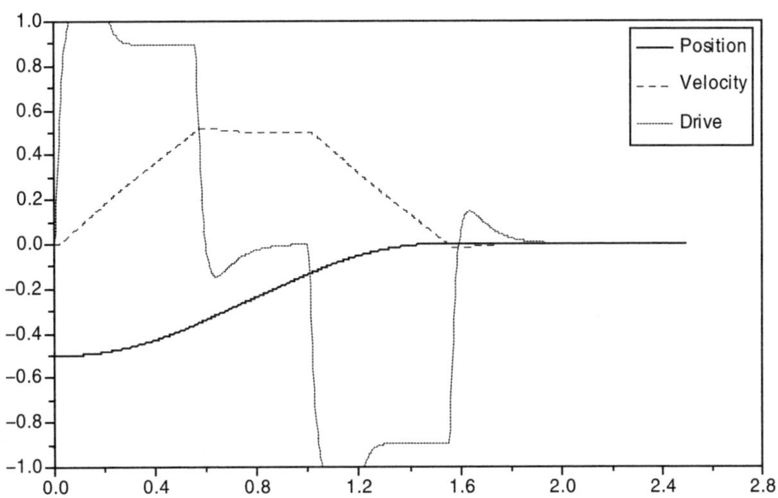

Figure 8.13 Motor system from Figure 8.9 with trapezoidal velocity profile

Controller State Limiting (Anti-Windup)

Using the position profiler described in (8.38) results in a system that works well when the system is free to work as expected, but it is fragile: all of the effects that caused the limit cycle seen in Figure 8.9 are still there, and the system could easily encounter conditions that will make this limit cycle come out of hiding to cause problems. One case would be if the system met an obstruction while executing its trajectory that caused the position command to overrun the actual position; an example of this is shown in Figure 8.14. This is the same system as simulated before, with the same velocity profile, but with some mechanical drag simulated in the period from 0.5 seconds to 0.75 seconds.

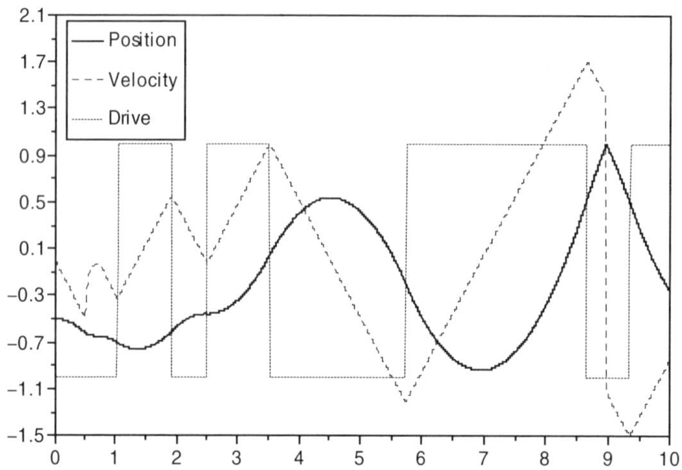

Figure 8.14 A trapezoidal velocity profile

What can be done? Recall that the system has three levels of integration: one in the controller and two in the motor. We can directly access the integrator in the controller. If we limit this integrator's state then when it is in its state-limited region, it will effectively stop being an integrator, and it will no longer contribute any phase shift to the system; such a change will mean that for large signals the phase shift will not be as bad.

The behavior of the integrator where its output is no longer making any difference is referred to as *windup*. Limiting integrator windup in such a manner is, of course, referred to as *anti-windup*. Just about any time you design an integrator into a control loop you should take measures to include some anti-windup, and you should consider just how much limiting is necessary and what form it should take.

There are a number of ways to limit an integrator's state. The two most popular are either simple integrator state limiting, or integrator output limiting. With integrator state limiting, the state of the integrator follows

$$x_n = \begin{cases} x_{min} & x_{n-1} + k_i u_n < x_{min} \\ x_{max} & x_{n-1} + k_i u_n > x_{max} \\ x_{n-1} + k_i u_n & otherwise \end{cases}. \tag{8.39}$$

This is simple, direct, and corrects most problems. I prefer to use integrator output limiting, which is slightly more complex but that will usually result in much less severe overshoot:

$$\begin{aligned} x_n &= \begin{cases} y_{min} - k_p u_n & x_{n-1} + \left(k_i + k_p\right) u_n < y_{min} \\ x_{max} - k_p u_n & x_{n-1} + \left(k_i + k_p\right) u_n > y_{max} \\ x_{n-1} + k_i u_n & otherwise \end{cases}. \\ y_n &= x_n + k_p u_n + differential\ term \end{aligned} \tag{8.40}$$

With this latter method the integrator state plus the proportional term of the system will never exceed the maximum system drive; not only will this prevent any excess integrator state from building up, it will also always give the differential term room in which to work.

Figure 8.15 shows the result of adding this anti-windup scheme into the system. While the settling time is nothing to write home about, the system does not launch itself into violent oscillations as a result of the initial fault. Using the "straight" anti-windup shown in (8.39) would have given quicker results at the cost of more overshoot.

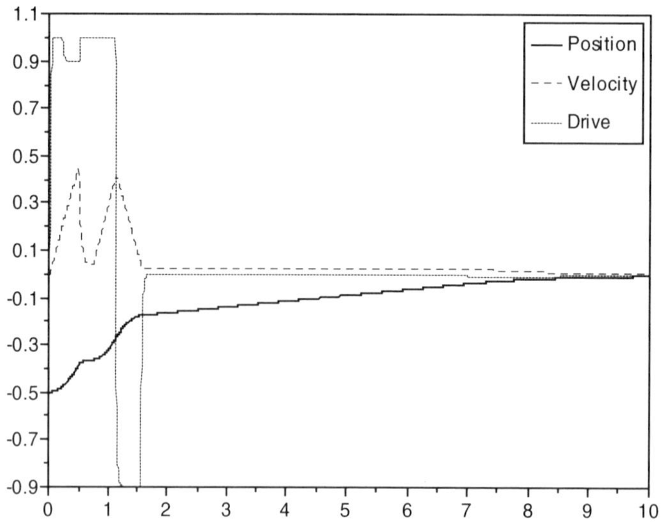

Figure 8.15 System of Figure 8.14 with velocity profile and integrator anti-windup

Plant State Limiting

Limiting the state of the plant is a much more indirect process than limiting the state of the controller; in the case of the controller, you have direct control of all of its features and you can pretty much do whatever you want with it. In the case of the plant, you cannot simply reach in and change the instantaneous value of a state, and in some cases you cannot even measure it directly. Nonetheless, you can frame a problem in terms of limiting the plant state and get advantageous results.

Figure 8.16 shows a control system that limits a plant state, in this case a motor's velocity. This is accomplished by arranging an outer position loop to generate a velocity command that is limited. This velocity command is applied to an inner velocity loop that uses a differentiator as a velocity sensor and a PI controller to hold the motor velocity at the commanded value. If you ignore the nonlinear behavior of the velocity command limiting, this technique is just a complicated way to implement straightforward PID control for the motor; with the velocity command limiting, however, this loop becomes a way to insure that the motor speed will not exceed some reasonable maximum.

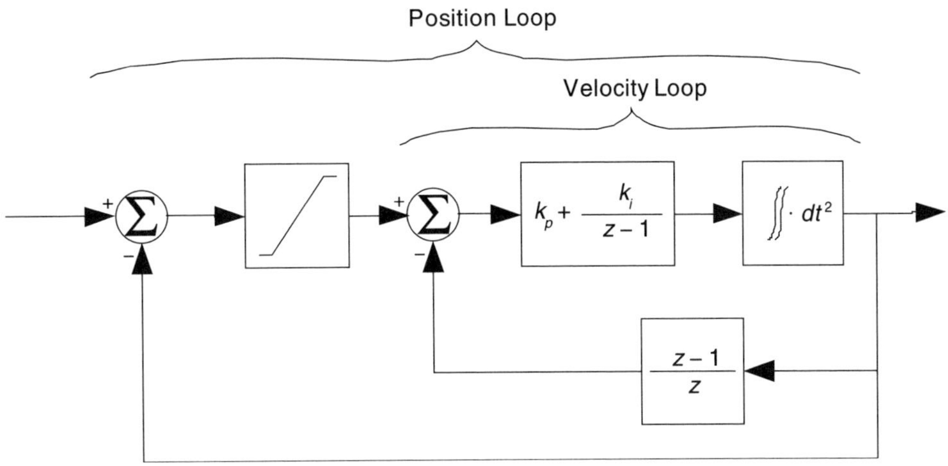

Figure 8.16 A two-loop system for limiting plant state

In general, efforts to limit plant states will be similar to this example. Because you cannot limit the plant state directly, you must arrange your controller to sense the relevant state and control around it.

Pulsating Drive

Rather than driving a plant with a continuous voltage or current, one can pulse the drive on and off, or between off and some lower level. The average drive is controlled by changing the duty cycle of the drive, either with pulse-width modulation (PWM) or by driving with constant-duration pulses and controlling how many of these pulses are applied.

In cases where the actuator can take the treatment, the driver circuitry can be simplified by pulsing the full actuator voltage on and off. Not only is this technique useful for overcoming plant nonlinearities, but it also makes for such a simple, effective and inexpensive drive that it is often employed just to reduce system cost.

Pulsating drive is very effective as a means to linearize the drive to heaters, in reducing the nonlinearities due to friction, and any other place where there is a nonlinearity that is sensitive to the amplitude of the applied drive.

In a heater with a resistive element being driven by a voltage (or current), the amount of heat generated is the square of the applied voltage divided by the heater resistance. Driving such a device with a continuous voltage results in a pretty severe nonlinearity. Pulsing the drive, however, results in a heat flow that is, on average, significantly more linear.[2] Pulsating drive is also very effective against friction.

Friction, as described in section 8.2, creates a nasty nonlinearity because the plant does not respond at all to small inputs, then suddenly starts responding—often with a bang. Attempts to deal with friction using linear control are nearly always doomed to failure; the best one can hope for in such a case is a small limit cycle that can be ignored or at least tolerated.

This is not necessary, however. One fairly easy and effective measure to reduce nonlinearities from friction is to apply a pulsating drive to the plant, usually PWM. With PWM drive, the actual drive (current or voltage) applied to the actuator is not a continuous function of the drive command. Rather, for small drive commands the actual drive is pulsed on and off with a duty cycle that provides the correct average drive. Some systems will transition to a linear drive at some point, while some will maintain the pulsed drive all the way up to the maximum drive command.

Note that to overcome friction or linearize a heater, the PWM drive to the plant should *not* occur at such high frequencies that the current through the actuator is smoothed. The switching in the PWM drive is not being done for efficiency as in the case of a switching power supply, and the effect is lost if it is too fast. PWM drive frequencies must be slow enough that the joint with friction moves a small but appreciable amount; with small motor drives they are generally in the 10Hz to 1000Hz range, where modern switching amplifiers may operate above 1MHz.

The advantage of a PWM drive to a plant that is limited by friction is that the plant will always move for small inputs. This means that the plant can be slowed down to a crawl without the jerkiness associated with a continuous drive. In fact, using PWM drive to a plant will sometimes mean that the controller doesn't have to have an integrator to achieve zero steady-state position error. Sometimes when you get lucky this way, you can drop from a PID loop to a PD loop.

---

2 Once you can stop worrying about the voltage-squared nonlinearity of your heater you can start worrying about the fact that resistance changes with temperature, and that heat flow itself is often a nonlinear phenomenon, and the next thing and the next...

There are disadvantages, however. The plant will always move by a discrete amount, however small. This small discrete movement means that you can never achieve truly zero error—and if you try to reduce this error by reducing the PWM on time the severity of the nonlinearity will go up. The velocity to drive relationship will still be quite nonlinear, although never as bad as without PWM. Finally, the energy in each drive pulse can be high; the actuator must be able to withstand the electrical and mechanical stresses generated by the PWM.

If the plant is at rest, the amount that it will move in response to a single pulse depends on the level of friction in the joint, the force generated by the actuator in response to a pulse, and the length of the pulse. This can be found by calculating the plant motion in response to a single pulse.

The plant will accelerate in response to a pulse, then it will slide to a stop. The following discussion assumes that the pulse width is at least as long as the actuator's electrical time constant, yet short enough that the plant velocity doesn't get large enough to create any restraining forces other than friction. Given that, the plant's velocity and position profile will follow the relationship

$$v\left(t_0 + t_{on}\right) = \int_0^{t_{on}} \frac{f_s - f_f}{m} dt = T_p \frac{f_s - f_f}{m} \tag{8.41}$$

$$x\left(t_0 + t_{on}\right) = x\left(t_0\right) + \iint_0 \frac{f_s - f_f}{m} dt^2 = x\left(t_o\right) + \frac{t_{on}^2}{2} \frac{f_s - f_f}{m} \tag{8.42}$$

where v is the plant velocity, x is the distance traveled by the plant, f_s is the force generated by the actuator, f_f is the friction force, m is the mass moved by the actuator and t_p is the pulse duration. After the pulse is removed, the motor will slide to a stop in time t_s, which can be found from

$$t_s = \frac{w_{on} m}{f_f} = t_{on} \frac{f_s - f_f}{f_f} \tag{8.43}$$

Where v_{on} is the plant speed attained in (8.41).

The total distance that the plant travels in response to a pulse can then be determined:

$$x(t_0 + t_{on} + t_s) = x(t_0) + \frac{t_{on}^2}{2}\frac{f_s - f_f}{m} + \frac{t_s^2}{2}\frac{f_f}{m}. \tag{8.44}$$

(8.43) can be substituted into (8.44) to eliminate t_s:

$$\begin{aligned} x(t_0 + t_{on} + t_s) &= x(t_0) + \frac{t_{on}^2}{2}\frac{f_s - f_f}{m} + \frac{t_{on}^2}{2}\frac{(f_s - f_f)^2}{f_f m} \\ x &\simeq x(t_0) + \frac{t_{on}^2}{2}\frac{f_s(f_s - f_f)}{f_f m} \end{aligned}. \tag{8.45}$$

Equation (8.45) indicates that the plant will always move in response to a pulse, but that the amount that the motor moves will be proportional to the square of the pulse width.

These calculations only hold if the plant stops completely between pulses. Once the pulses are spaced closely enough that the motor does not stop between pulses, the system will start to act like a linear system with a constant force applied against the direction of motion and a driving force equal to the duty cycle times the on-time force. The duty cycle at which this change in behavior appears is usually constant over a fairly large range of pulse on-times, and is equal to the ratio of the plant friction force divided by the drive force available from the actuator:

$$\rho = \frac{t_{on}}{t_s + t_{on}} = \frac{t_{on}}{t_{on} + \frac{f_s - f_f}{f_f} t_{on}} = \frac{f_f}{f_s} \tag{8.46}$$

Example 8.1 Setting PWM Duration

You are designing a motor controller for a motor with substantial friction. The motor has an armature resistance of 1.7 ohms, a torque constant of 5.9 mNm/A (millinewton-meter/amp), a rotor inertia of 3.8 gram-cm^2 and a rotor inductance of 110microhenry. The motor is attached to a mechanism with a moment of inertia of 5 gram-cm^2, a breakaway friction torque that can range from 2 to 5 mNm, and a running friction torque that can range from 1 to 2 mNm. You are driving this mechanism from a controlled-voltage source that can deliver from –5V to +5V.

- Find the minimum voltage needed to guarantee that the motor will always move. Set a running voltage that gives you a 20% margin over this value.
- Find the pulse on-time (t_{on}) that is required to limit the motor travel to no more than 5 degrees at a time.
- Find the motor travel for the above pulse time with the maximum expected friction.
- Compare this pulse on time with the motor's electrical time constant.
- Find the duty cycle where the motor no longer stops between pulses.

The torque at the motor armature is equal to the armature current times the torque constant. To generate 5mNm this current must be:

$$I_{min} = \frac{5\text{mNm}}{5.9\dfrac{\text{mNm}}{\text{A}}} = 0.85\text{A} .$$

The voltage required to develop this current across the motor armature can be found from the motor armature resistance:

$$V_{min} = I_{min} R_a = (0.85\text{A})(1.7\Omega) = 1.45\text{V} .$$

- To get a 20% safety factor the voltage drive would need to be V_d=1.74V. The motor torque at this voltage, when the motor is not running, is

$$T_s = \frac{k_m V_d}{R_a} = \frac{\left(5.9\frac{\text{mNm}}{\text{A}}\right)(1.74\text{V})}{1.7\Omega} = 6\text{mNm},$$

which is 20% greater than the specified 5mNm.

We can find the displacement vs. time from (8.45):

$\Delta\theta = \frac{t_{on}^2}{2}\frac{T_s(T_s - T_f)}{T_f J_m}$, therefore $t_{on} = \sqrt{\frac{2T_f J_m \Delta\theta}{T_s(T_s - T_f)}}$ (note that we are using torques, denoted with a T, instead of forces, denoted with an f).

To find the maximum allowable time we need to use the maximum displacement and the minimum T_f. This time is:

- $$t_p = \sqrt{\frac{2(1\text{mNm})(8.8\text{gcm}^2)(.095\text{rad})}{(6\text{mNm})(6\text{mNm} - 1\text{mNm})}} = 2.36\text{ms}\cdot$$

Given this time, with the maximum friction the displacement will be

- $$\Delta\theta = \frac{(2.36\text{ms})^2}{2}\frac{(6\text{mNm})(6\text{mNm} - 2\text{mNm})}{(2\text{mNm})(8.8\text{gcm}^2)} = .036\text{rad} = 2.1\text{degrees}\cdot$$

The electrical time constant of the motor is simply the L/R constant of the armature. This is

- $\tau = \frac{110\text{mNm}\mu\text{H}}{1.7\omega} = 65\mu\text{s}$, which is much smaller than t_p.

The duty cycle where the motor will start running continuously ranges from

- $$\rho_{min} = \frac{T_f}{T_s} = \frac{1\text{mNm}}{6\text{mNm}} = 0.17 \text{ to } \rho_{max} = \frac{T_f}{T_s} = \frac{2\text{mNm}}{6\text{mNm}} = 0.33.$$

I have not given you on overall view of how the plant will act once the duty cycle for continuous rotation given in (8.46) is exceeded. This is because the plant behavior varies at drives above this duty cycle. If the plant were actuated by a DC motor with a controlled-current drive, its speed would be unbounded above this duty cycle, as the motor would integrate the extra force above the friction force. If the plant were actuated by that same DC motor but with a controlled voltage drive, the motor's velocity would rise to some equilibrium value and stay steady. Finding the plant behavior for higher duty cycles remains a task for the control system designer.

Deadband

Some plants, in particular plants being driven with pulsating drive and plants which have backlash, cannot be practically driven to have zero error. The inevitable residual error can cause the system to hunt around the equilibrium point, particularly in systems with PID controllers, where the integrator will insist on increasing its output until something happens.

If there is nothing that can be done to prevent the residual error, it will be necessary to find a way to cope with it. The method that works best for this is to put some deadband in the controller's error term. Deadband sets the drive to the motor to zero when the input error is within some defined limit, so the motor will come to rest even if it is at a slightly incorrect position. The two most commonly used deadband methods are defined in (8.47) and (8.48); they are shown graphically in Figure 8.17.

$$u_{out} = \begin{cases} u_i + u_d & u_i < -u_d \\ 0 & -u_d \le u_i \le u_d \\ u_i - u_d & u_i > u_d \end{cases} \tag{8.47}$$

$$u_{out} = \begin{cases} u_i & u_i < -u_d \\ 0 & -u_d \le u_i \le u_d \\ u_i & u_i > u_d \end{cases} \tag{8.48}$$

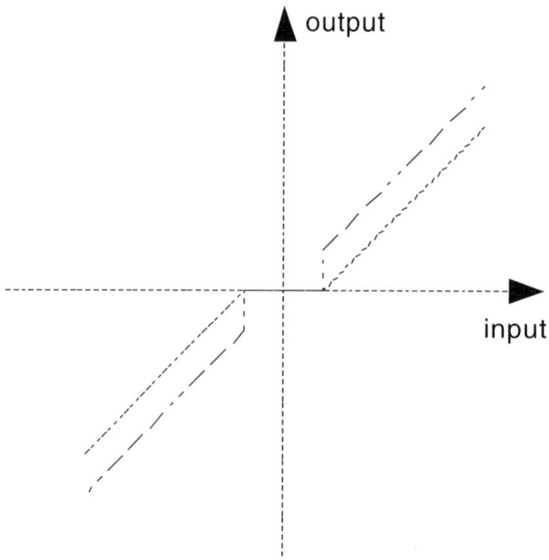

Figure 8.17 Deadband input/output relationship

The simplest way to apply deadband is to simply insert a deadband operation between the summing junction of the control loop and the rest of the controller. This will suffice in most cases, and where deadband is necessary it will be a distinct improvement over a control loop with no deadband at all.

Figure 8.18 shows the preferred method of designing a controller for a plant with backlash. In this scheme, the deadband is only applied to the proportional feedback—derivative feedback is treated separately, so the motor velocity will always have a damping effect on the motor motion, and so there won't be any "hiccups" in the motor drive if the error term passes through zero.

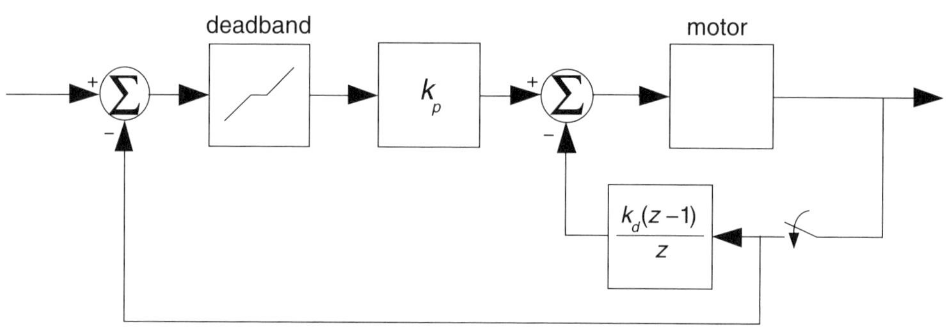

Figure 8.18 Deadband in a PD controller

If the system is to be designed for PID control, then the proportional gain block in Figure 18 should be replaced with a PI block, and the differential block (if used) left where it is.

Using deadband in a control loop with a PID controller can present some interesting difficulties, and should be approached with care. When the loop error is within the deadband, the effective loop gain goes to zero. Because a PID driving a motor is only marginally stable as the loop gain decreases, this can cause problems, which are compounded by using PWM. Additionally, the deadband can leave the integrator with some nonzero stored value that will waste power and heat the motor.

Setting the amount of deadband in a system can be problematic. At a minimum, the deadband should be set to the amount that the motor can jump if the error is just outside the deadband region, to prevent small limit cycles. For a mechanism that has significant friction and backlash, however, this deadband will prove to be too small in practice. Normal system design will usually include quite a bit of cut-and-try tuning of deadband parameters to arrive at acceptable system behavior.

Change the Plant

Even for a control engineer who is equipped with a big bag of tricks to deal with system nonlinearities, there are some systems that cannot be adequately controlled because of their nonlinear behavior. Some plant nonlinearities are simply incompatible with system goals. Plant uncertainties will limit the bandwidth that can be achieved, actuator saturation will limit the final speeds that can be achieved, and friction, backlash and cogging will limit the final accuracy of your system. Always be sensitive to the fact that there are times when a plant cannot be controlled to reach desired objectives, or the engineering effort required to design a controller will be expensive. In such cases it may be most profitable, or even essential, to change the plant rather than the control system.

8.5 Conclusion

This chapter has barely scratched the surface of nonlinear control theory; one could fill a room with books about nonlinear control and still not have a complete treatment to the subject—particularly because it is still an ongoing field of study in universities.

Even though nonlinear control *is* a subject of ongoing studies, it has the advantage that in the field it is often most useful to apply common sense and a knowledge of a few methods to a problem—strictly mathematical approaches are often not feasible, even for an expert, and significant performance gains can often be had just by taking measures that "feel right."

Finally, it is important to remember that *all* control problems are problems in nonlinear control. If you find yourself tearing your hair out because your system simply doesn't seem to be behaving the way the linear analysis says that it should, you should ask yourself if the system is "linear enough" for the analysis to be applicable.

9

Measuring Frequency Response

9.1 Overview

When you design a control system using any of the frequency response methods—Bode plot, Nyquist plot or Nichols chart—it is not necessary to refer to the z-domain transfer function of the plant or controller, except to find the system gain and phase response over the frequency range of interest. It is possible, therefore, to use a set of frequency response data without having an exact z-domain transfer function in hand.

This freedom from needing an exact transfer function means that not only can you measure your linear system behavior's frequency response, but you can measure a *nonlinear* system's frequency response at a certain amplitude and use describing function analysis to do your design. In the end, you can do a very good job of control system design without ever having a known model of your plant; it is sufficient to have a set of measured frequency responses complete enough for your design.

Knowing this, one is immediately motivated to ask, "So, how do I measure the frequency response?" There are measuring instruments that have been developed to do this,[1] but they are expensive and in a digital control system they are difficult to integrate. For little more than the effort required to integrate a control systems analyzer into your system, you can build the necessary interfaces into your software to allow you to collect and analyze the necessary data on your PC, with the added bonus that the resulting data will be in a form that you can immediately use for designing your control system.

---

1 For example, the Agilent™ HP3563A.

There are a number of facets to measuring frequency response that must be addressed. You must excite the system with the correct sinusoidal waveform, and you must collect the resulting system behavior and extract the relevant frequency response data from it. This must be done in a manner that is relatively immune from the noise than inevitably permeates a real-world control system, that is practical and easy to use, and doesn't add significantly to the cost of your system.

9.2 Measuring in Isolation

The most direct way to measure a system's response at a given frequency is accomplished by driving the system under test with a sine wave at the desired frequency while monitoring the relevant system output(s). Then find the amplitude and phase of the response, and compare this amplitude and phase to the amplitude and phase of the injected sine wave. To find a complete system frequency response over some span, you excite the system with a sine wave, taking a set of measurements at a frequency, stepping the frequency up (or down) and repeating as necessary until the desired data set has been collected.

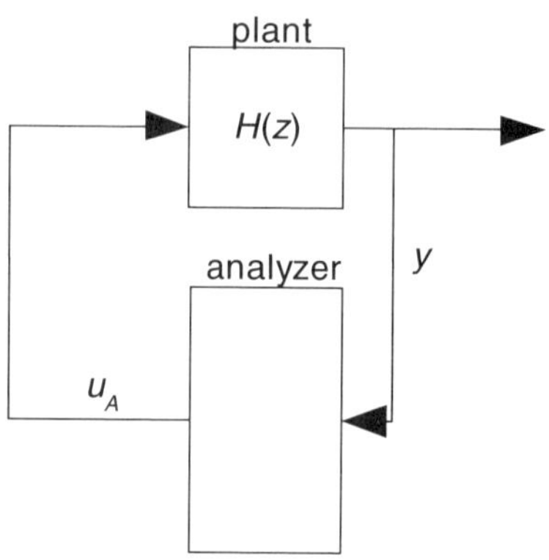

Figure 9.1 Measuring frequency response

Take the setup shown in Figure 9.1, where the goal is to find the frequency response of the plant whose transfer function is $H(z)$. For any given frequency f (in cycles per sample) we set the input signal

$$u_A(k) = A_T \sin(2\pi f k) \tag{9.1}$$

where k is the sample index and A_T is chosen to protect us from overloading the plant. Assuming that the plant is stable and linear, the output of the plant will be of the form

$$y(k) = A_T A(f)\sin\left(2\pi fk + \phi(f)\right) + y_T(k) \tag{9.2}$$

where $A(f)$ and $\phi(f)$ are the gain and phase of the frequency response at f and $y_T(k)$ is a transient signal which will go to zero as k goes to infinity.

To actually find the values of $A(f)$ and $\phi(f)$ you can measure the response over some finite number of samples and compute the first term of the Fourier series for $y(k)$:

$$\begin{aligned} A_I(f) &= \frac{2}{N}\sum_{k=0}^{N-1} y(k)\sin(2\pi fk) \\ A_Q(f) &= -\frac{2}{N}\sum_{k=0}^{N-1} y(k)\cos(2\pi fk) \end{aligned} \tag{9.3}$$

and

$$A = \frac{\sqrt{A_I^2 + A_Q^2}}{A_T} \tag{9.4}$$

$$\phi = \begin{cases} \tan^{-1}\left(\dfrac{A_I}{A_Q}\right) & A_Q \geq 0 \\ \tan^{-1}\left(\dfrac{A_I}{A_Q}\right) + 180^\circ & A_Q < 0 \end{cases} . \tag{9.5}$$

The equations in (9.3) only work well if the samples are taken over an integer number of cycles of the input, in other words if N is an integer multiple of $1/f$, and they work best if the transient signal is reduced as much as possible. You can minimize the transient by starting your excitation at a point where the sine wave is zero, by switching frequencies at a zero-crossing point, and by allowing the transient to die out before collecting data.

If you recall Euler's identity,

$$e^{j\theta} = \cos\theta + j\sin\theta, \tag{9.6}$$

(9.3) and (9.4) become

$$A(f)e^{j\phi(f)} = \frac{2}{jA_T N}\sum_{k=0}^{N-1} y(k)e^{j(2\pi fk)}, \tag{9.7}$$

where the $\pi/2$ term on the right-hand side compensates for the fact that the excitation is a sine wave rather than a cosine wave. This is a handy relationship to remember if you are using a math package to do your computation, as you can reduce (9.6) down to a single vector multiply with very efficient computation using such a package.

For example Figure 9.2 shows the response of a second-order system at a frequency of $f = 1/500$. There is an initial turn-on transient, but the system is already settling out by the end of two cycles. If we apply (9.3) to the entire data sample in Figure 9.2 we see an error of about 20% from the actual transfer function, but applying it to the last 500 samples gives us an error under 3%.

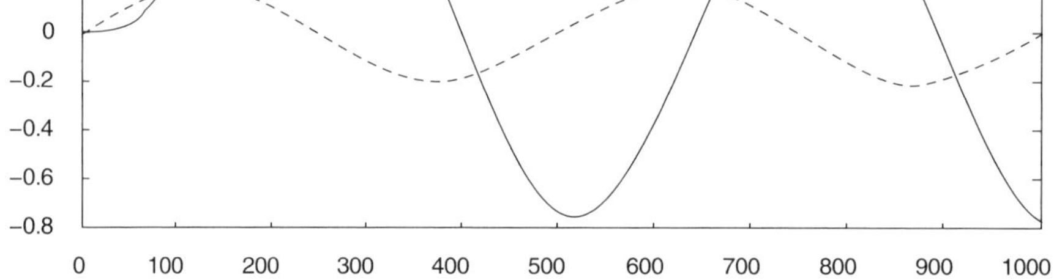

Figure 9.2 Example time-domain input and output

Normally, you are going to be interested in making measurements at a number of closely spaced frequencies. In this case, the best way to reduce the startup transient is just to ensure that the input sinusoid changes smoothly—as long as it does not stop or exhibit phase jumps at the frequency boundaries, then the transients will be negligible for most systems.

9.3 In-Loop Measurement

The setup in Figure 9.1 and equation (9.3) are sufficient for taking measurements of a well-behaved system in isolation. If, however, you are interested in more than just the behavior of one system (or subsystem) taken in isolation, if you want to know how your system behaves in closed loop, or if you are dealing with an unruly or unstable plant it falls short.

What you need in such a case is a setup and analysis method that lets you measure the frequency response of a portion of a working system without significantly rearranging the system structure—specifically, without opening up any working control loops.

This setup should allow you to measure the frequency response of a portion of the system without opening the control loop. To do this, note that the transfer function of a block is the ratio of its input to its output. In the discussion so far, we've excited a block with a sine wave and extracted its output—but in the process of extracting the output parameters in (9.9) we divided by the first Fourier term of the sine wave excitation; that is, $j\,A_T$. If we choose our block, ensure that it has sinusoidal input and output at our desired frequency, then measure both its input and its output and divide their first Fourier term outputs, then we will have the block's gain and phase at the frequency in question.

Taken from a purely signal-flow perspective, the setup to measure frequency response is straightforward: place a summing junction in your system at a convenient location, then monitor the signals that sit at the input and output of the section of the system whose response you want to measure. The summing junction allows us to inject a sine wave which will then pervade the entire system, while the two outputs will let us get the pair of numbers that we need in order to determine the ratio of a block's input to its output.

Figure 9.3 shows the signal paths for measuring the response of the plant (note that any drivers, DACs, ADCs and sampling are lumped into the plant model; we're only interested in the plant behavior as seen by software). A signal, u_A is injected into the system right before the plant; the drive to the plant is picked off at the reference, u_R, and the system error is picked off at y. Assuming that the system command is held steady, this setup will measure just the plant response.

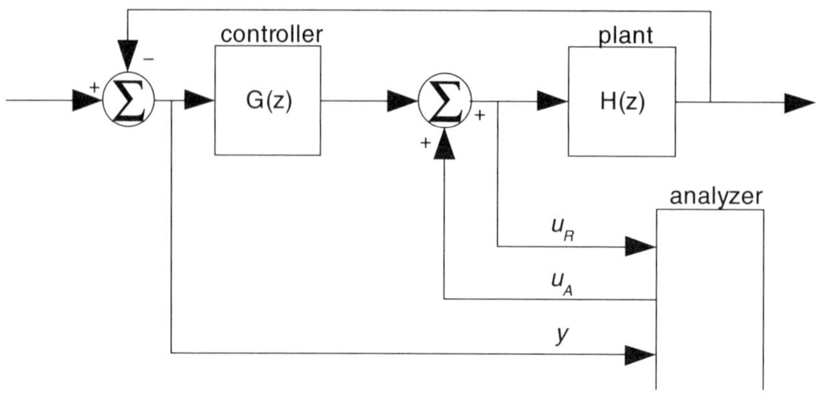

Figure 9.3 Setup for measuring plant response

Figure 9.4 shows the signal paths for measuring the open-loop response of the system. Here the signal is injected in the same place, and the reference pickoff is still immediately before the drive signal, but the output pickoff is measured after the signal has passed through both plant and controller.

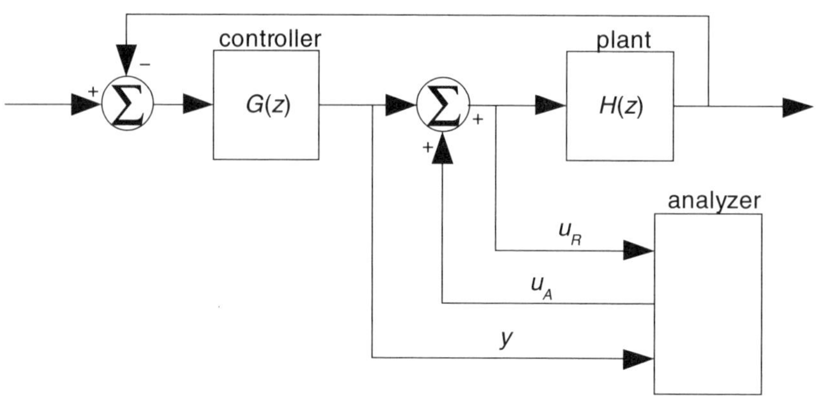

Figure 9.4 Setup for measuring open-loop response

The way to extract the frequency response of the subsystem under test is the same for Figure 9.3 and Figure 9.4: inject a swept-frequency sine wave at u_A at each frequency point, apply (9.3) to u_R and y, and divide the two resulting numbers—the result is the frequency response at that point.

This is done with the following set of equations. First, find the first Fourier coefficient for the two variables:

$$U_R(f) = \frac{2}{N}\sum_{k=0}^{N-1} u_R(k)\, e^{j\left(2\pi fk - \frac{\pi}{2}\right)}$$
$$Y_1(f) = \frac{2}{N}\sum_{k=0}^{N-1} y(k)\, e^{j\left(2\pi fk - \frac{\pi}{2}\right)} . \qquad (9.8)$$

Then divide the resulting complex numbers to find the value of the frequency response:

$$H(f) = \frac{Y_1(f)}{U_R(f)} . \qquad (9.9)$$

For example, say we wish to measure the plant response, open-loop response and closed-loop response of a system such as the one shown in Figure 9.3 and Figure 9.4, where we're sampling at 1kHz. We don't know it, but the plant transfer function is

$$H(z) = \frac{0.001z}{(z-1)^2} . \qquad (9.10)$$

We *do* know the controller transfer function, which is

$$G(z) = k_d \frac{z-1}{z} + k_p + k_i \frac{z}{z-1} \qquad (9.11)$$

with $k_d = 100$, $k_p = 2$ and $k_i = 0.01$.

With a frequency sweep from 0.1Hz to 500Hz, we get the plant input and output, which are shown in Figure 9.5 and Figure 9.6. You can see that as the magnitude of the response gets small the signal gets noisy, as seen on the right in Figure 9.5 and on the left in Figure 9.6.

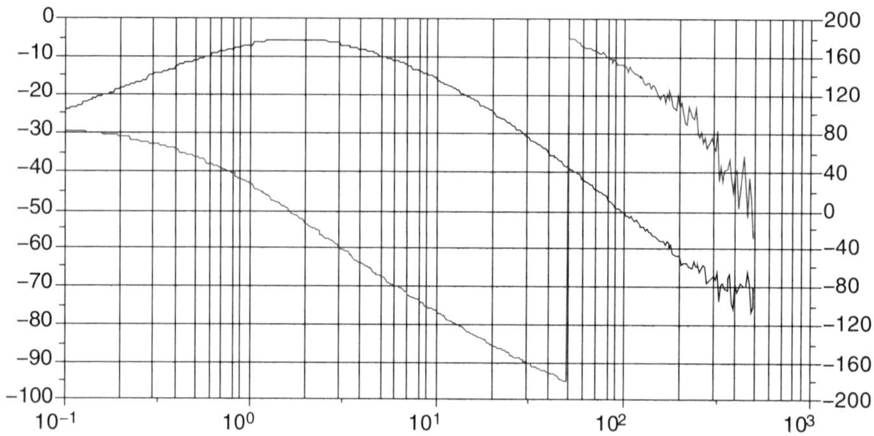

Figure 9.5 Measured Bode plot of the plant output

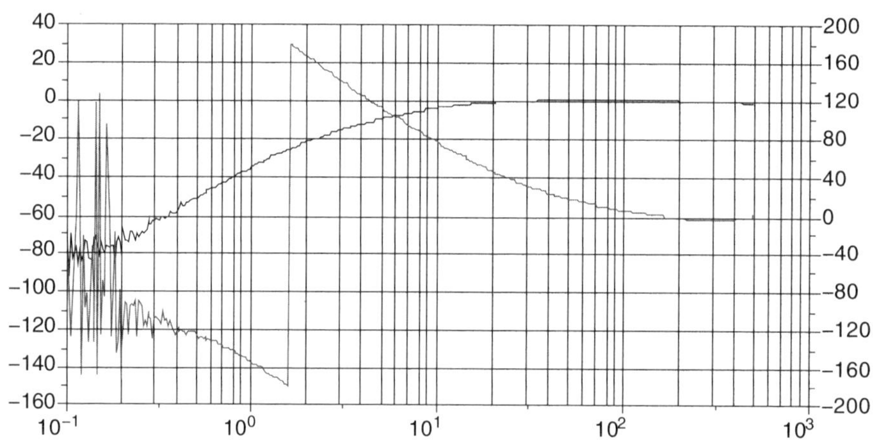

Figure 9.6 Measured Bode plot of plant input

Taking these two sets of data and performing complex division on each frequency point, then plotting the results yields the response shown in Figure 9.7. This measured plant response can then be used with a controller model (which should certainly be exact!) to tune the system.

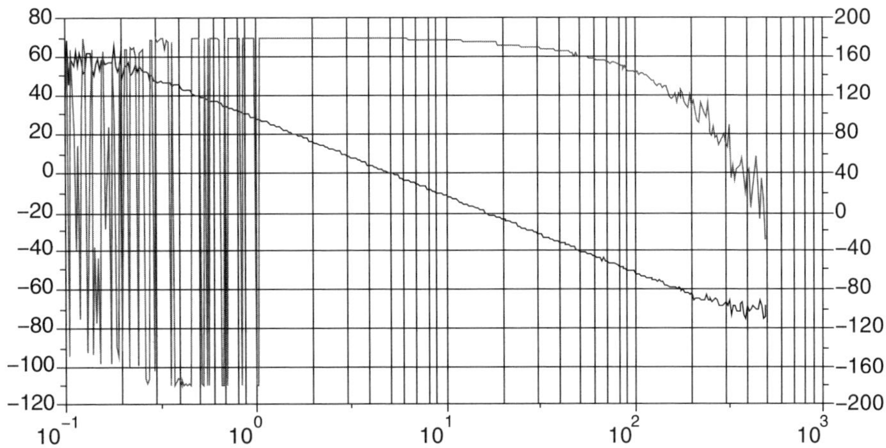

Figure 9.7 The measured plant response

Rearranging the data collection to that shown in Figure 9.4, we measure the output of the controller and the input to the plant. The input to the plant hasn't changed, and the measured controller output is shown in Figure 9.8 (which is, incidentally, the closed-loop response multiplied by −1). Dividing the controller output by the plant input gives us Figure 9.9.

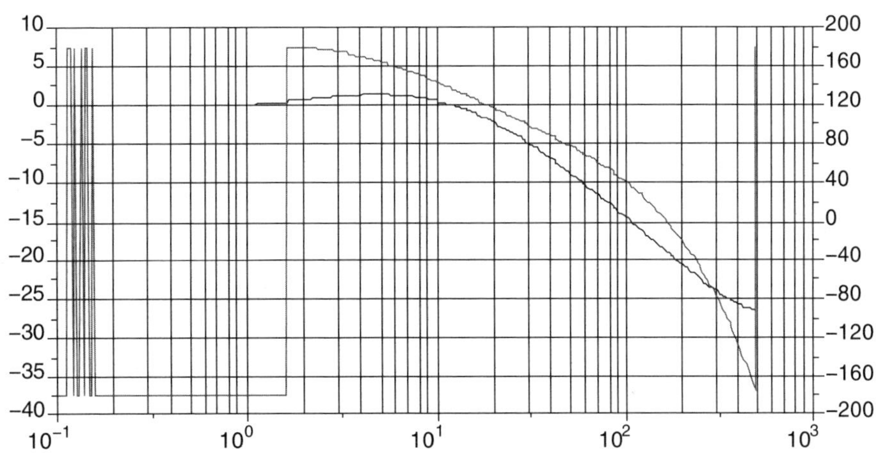

Figure 9.8 Measured controller output (closed-loop response)

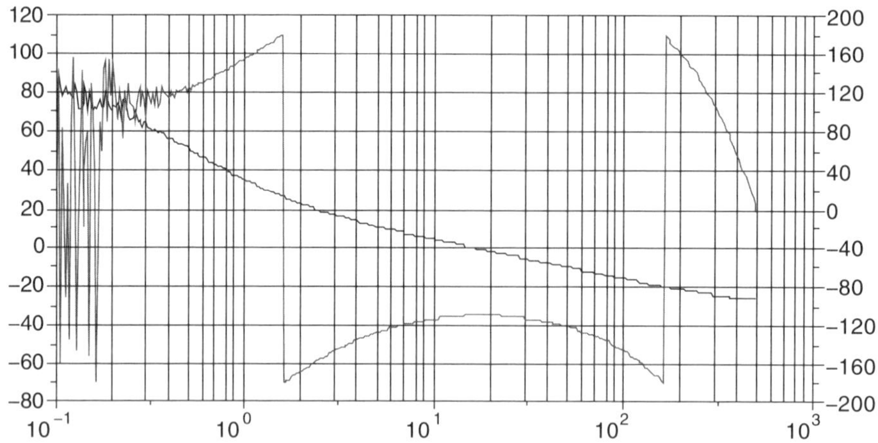

Figure 9.9 The measured open-loop response

On inspecting Figure 9.9, it can be immediately seen that the gain-crossing frequency occurs at about 16Hz with a phase margin of about 68°, and that there are phase-crossing frequencies at 1.6Hz and 160Hz, with gain margins in excess of 20dB.

9.4 Real-World Issues

Noise

Noise is an inevitable part of any measurement, and in control system design it is often useful to get frequency response data over several decades of the plant response. Unfortunately, signals can get very small at the frequency extremes and the noise level is often fixed. When doing these measurements, noisy data is inevitable, but as long as you understand the effects of the noise on your measurement, you can mitigate them.

In Figure 9.7, we saw the effect of measurement noise at both ends of the spectrum. The measurement noise is constant across the frequency spectrum, but the quantity being measured varies: at the low-frequency end of the spectrum the loop gain is very high and the closed-loop gain is unity. This causes the signal at the input to the summing junction to be very small, so it can be corrupted by noise. At the high-frequency end of the spectrum the loop gain is small, so the signal level at the output of the summing junction is small and once again the measurement noise can dominate.

To mitigate the effects of noise in the measurements, you must increase the signal to noise ratio in the final numbers arrived at using (9.3). This can be done in one of two ways: first, increase the signal by increasing the amplitude of the sine wave that you insert into your system; and second, increase the amount of time that you collect data at any particular frequency.

Increasing the signal amplitude gives the obvious advantage of more signal. You must take care, however, that you are not increasing the drive signal to the point where your system goes nonlinear through overflow or other effect (see the next section).

Increasing the amount of time over which you collect data gives you a better signal to noise ratio because the operation of (9.3) gives a result that is coherent (i.e., synchronized) with the input sine wave, but the noise signal is incoherent. As a result, the average signal amplitude rises in proportion to the length of time you collect data, but the expected amplitude of the noise only rises as the square root of this time—so increasing the collection time at each frequency by a factor of 4 will double the eventual signal to noise ratio. I have found through experience that taking a minimum of 1000 points per frequency will usually generate good data.

Nonlinearities

Nonlinearities are an inevitable part of designing a control system, because ultimately there are no physical systems (including real software running on real processors) that don't exhibit nonlinearities. Often, the nonlinearities that we must design around are slight compared to our control system requirements; in such cases it is not a bad idea to design our system using describing function analysis—and swept-sine frequency response data is exactly what is necessary for such design.

In our original analysis we justified using (9.3) to demodulate our data because of the form observed in (9.2). But (9.2) only follows from (9.1) if you assume that the system being tested is linear, or is being operated in a linear region. With a system that is *not* operating in a linear region, (9.2) is not, in general, valid. In fact, the whole notion of using frequency response analysis is not generally valid for nonlinear systems.

What to do? Is this entire methodology one that only applies to that small subset of systems that is acting exactly like a linear system for the particular input that we expect? Fortunately, the answer to this question is no. In the case when we are testing a stable nonlinear system, then the response to a sinusoidal input is still periodic but in addition to containing energy at the fundamental frequency *f* it will also contain energy at DC (*f* = 0), 2*f*, 3*f*, and so on, so (9.2) becomes

$$y(k) = y_T(k) + A_T \sum_{n=0}^{\infty} A_n \sin(2n\pi fk + \phi_k) \tag{9.12}$$

and A_1 (the fundamental Fourier term) is the same as A in (9.3). If the value of A_1 is significantly larger than any of the higher-order responses (A_2 through A_∞) then describing function analysis says that we can conveniently ignore the nonlinearities and just use the system as described by A_1(f) and $\phi_1(f)$.

Fortunately, the demodulation described in (9.9) or (9.3) will only respond to the fundamental energy in $y(k)$, so the DC component (A_0) and all of the higher-order terms in the summation evaluate to zero. What emerges from these demodulation operations are the amplitude and phase information for the fundamental frequency only—which is usually exactly what you want in describing function analysis.

But how can you tell if the system you are measuring is operating in its nonlinear region, and how can you quantify "how nonlinear" it is? The three methods that you can use are: One, look at your system's behavior to see if an output, input, or intermediate value is hitting a maximum limit during operation; two, repeat sweeps with different input amplitudes and compare their results and three, use measures of data correlation.

Observing your system behavior is probably the best way to ensure that the system is operating in its linear region if you have a good idea of how your system works. Examples of parameters to inspect are: the system drive to see flat-topping on its output, the mechanisms themselves to see if they are hitting any stops, the system sensors to see if they are saturating even when the thing they are measuring isn't, and finally making sure that no signals are so small that they are falling into the cracks of gear train backlash or ADC/DAC precision.

To tell if a system is in a linear region, you should repeat your sweeps with several different input amplitudes and compare the results. If the results are substantially the same, then chances are high that you are operating in a linear region. If, however, the results are different, then you have a pretty good indication that you have left the linear region of operation for your system.

Repeating sweeps with different drive values will not only tell you whether you are taking data on a system that is in a nonlinear domain, it will also give you some guidance as to whether your system will be stable with that given input magnitude.

To use correlation to gauge system linearity, you need to observe that the only term measured in (9.12) that is actually used is the fundamental term. If you can estimate the extent of the higher-order terms, then you can get an idea of how much your nice pretty sine wave is being mangled as it passes through the system. So, you could express your correlation with the equation

$$\rho = \frac{A_1}{\sum_{k=1}^{\infty} A_k} \tag{9.13}$$

where $\rho = 1$ indicates a perfectly linear system, and lesser values of ρ indicate a "less linear" system.

But how do we extract this result from measured data? This can be done by observing that if you remove the DC component from your signal, square the result and average you get:

$$\frac{1}{N}\sum_{k=0}^{N-1}\left(y(k)-\overline{y(k)}\right)^2 = \frac{A_T^2}{2}\sum_{k=1}^{\infty} A_k^2 . \tag{9.14}$$

Now recall that from the results of (9.3) you can extract the value for just the fundamental:

$$A_Q^2 + A_I^2 = A_T^2 A_1^2 . \tag{9.15}$$

So, you can find your correlation from your measured results:

$$\rho = 2\frac{\left(A_Q^2 + A_I^2\right)}{\frac{1}{N}\sum_{k=0}^{N-1}\left(y(k) - \overline{y(k)}\right)^2}. \tag{9.16}$$

9.5 Software

How do you integrate this all into your system? The answer to that question depends a lot on how your particular system is architected. Specifically, how your sampling rate relates to your external communications bandwidth and real-time capabilities, and how much extra processor bandwidth you have at your disposal, determines how much functionality you put where.

I will give an example set of software that you might use if your control system sports a communication link fast enough to send and receive each analyzer sample in its raw form at the desired sampling rate.

The entire system is shown in Figure 9.10. The block marked "analyzer" is implemented on the host, and communicates to the embedded processor via the communications link. The dotted boxes marked "tap" enclose the summing junctions and switches that connect one tap or another to the analyzer. The block marked "controller" is the controller being developed, and the block marked "plant" is the system being controlled.

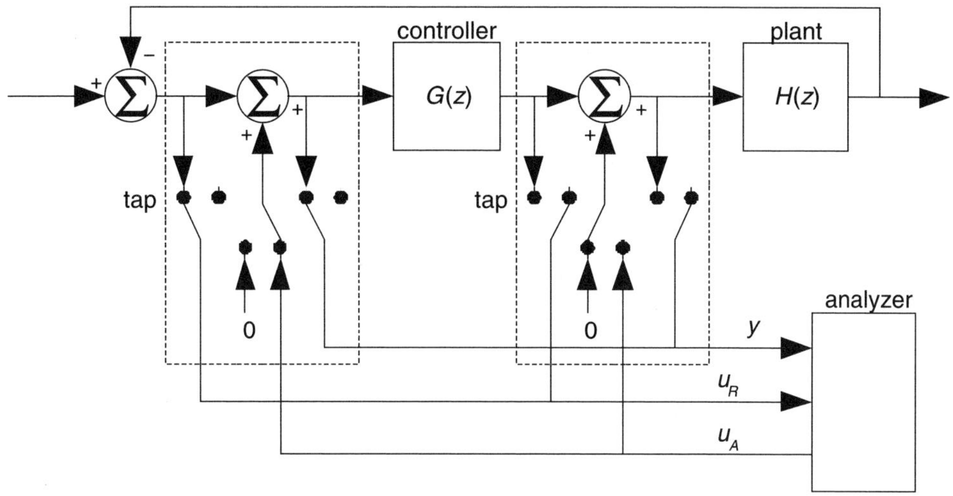

Figure 9.10 Setup for measuring plant response

Code

Assuming that there is a communications link, the additional software consists of three parts: the host-based stimulus generator, the embedded summing junctions, and the host-based results parser.

Embedded Code

Listing 9.1 shows the generic portion of the analyzer code that must be embedded in the system. This code depends on having three functions from the data interface: analyzerDataAvailable must return true if there is data to be read by analyzerGetInput; analyzerGetInput must return the latest stimulus data point sent by the host; and analyzerOutput must send the given output values to the host.

```
static int analyzerOut1, analyzerOut2;    // Output signals
static int analyzerIn;                    // Input signals
static bool inCurrent = false;
typedef enum analyzerTapEnum
       {NONE = 0, IN = 1, OUT1 = 2, OUT2 = 4} analyzerTapEnum;

void analyzerUpdate(void)
{
  if (inCurrent)
  {
    // Send the outputs to the host.
    analyzerOutput(analyzerOut1, analyzerOut2);
  }
  inCurrent = analyzerDataAvailable();
  if (inCurrent)
  {
    analyzerIn = analyzerGetInput();
  }
}

int analyzerTap(int input, enum tapType type)
{
  if (inCurrent)
  {
    if (type & OUT1)  analyzerOut1 = input;
    if (type & IN)    input += analyzerIn;
    if (type & OUT2)  analyzerOut2 = input;
  }

  return input;
}
```

Listing 9.1 The analyzer code

The function analyzerTap implements the taps shown in Figure 9.10. Depending on the value of the "type" input it will set the analyzerOut variables and inject the analizerIn signal into the summing junction. The function analyzerUpdate implements the interface between the host and the embedded system. It ensures that there is data waiting, receives it, and sends the most recent collected analyzer data to the host.

Listing 9.2 shows how the analyzer code is used. The normal operation of this code is to get the ADC input, update the controller, and send the controller's drive command to the DAC. To implement the analyzer, we insert the two calls to analyzerTap and the call to analyzerUpdate.

```
analyzerTapEnum inputTap, outputTap;
...

  adcIn = getAdcInput();                 // Get the input value
  adcIn = analyzerTap(adcIn, inputTap); // perform the tap

  drive = controllerUpdate(adcIn);       // update the
controller
  drive = analyzerTap(drive, outputTap);// perform the tap

  setDacOutput(drive);                   // update the DAC
  analyzerUpdate();                      // update the analyzer
```

Listing 9.2 Using the analyzer dode

The values of the analyzer type inputs must be managed to ensure that each analyzer signal is only connected to one tap. How the values of inputTap and outputTap are set determines the measurement that will be taken. For instance, to implement the plant transfer function measurement shown in Figure 9.3, you would set inputTap = OUT1 and outputTap = IN | OUT2; to implement the loop transfer function measurement shown in Figure 9.4, you would set inputTap = NONE and outputTap = IN | OUT1 | OUT2.

Host-Side Code

The embedded code doesn't have much functionality: it just shoves the signals around. In order to make the system work, it is necessary to put most of the "brains" into the host side. The host side must generate the appropriate waveforms at the appropriate frequencies for the appropriate amount of time, and it must be able to receive the data, parse it, and show results.

Listing 9.3 gives the code that transfers the sine wave into standard output. It is presumed that this output will be piped to a file and sent to the embedded system using a terminal program or other means. The code generates a wave that is swept in frequency in a number of distinct segments, with the frequency varying logarithmically. The frequency of each segment and the segment length are calculated so the segment is an exact integer number of cycles long, with segments padded out in length as necessary to get a full cycle.

```
/***********************************************************************
  name: generateSine

  description:  Generates a sine wave with the correct characteristics

  inputs:
  startF:      The start frequency (in fractional samples) of the sweep
  stopF:       The end frequency (in fractional samples) of the sweep
  minSamp:     The minimum number of samples per frequency step
  numStep:     The number of frequency steps
  amplitude:   The amplitude of the output
  round:       The number to round the output to

  outputs:
  none, but generates output on cout
***********************************************************************/
```

Listing 9.3 The sine generation code (continued on next page)

```
void generateSine(double startF, double stopF,
                  unsigned minSamp, unsigned numStep,
                  double amplitude, double round)
{
  double  logStartF = log(startF),
          logStopF  = log(stopF);

  double phase = 0.0;   // The output phase, so it can be kept continuous
  for (unsigned step = 0; step < numStep; step++)
  {
    // Step through the frequencies one by one until done
    double freq;
    unsigned cycles, samples;

    // Calculate the nominal frequency
    if (numStep < 2) freq = startF;
    else freq = exp(logStartF + step * (logStopF - logStartF) /
                                          (numStep - 1));

    // This block of statements calculates the constants necessary for the
    // output at this frequency to be an integer number of cycles with no
    // less than minSamp samples at the frequency.
    cycles  = (unsigned)ceil(freq * minSamp); // Number of cycles
    samples = (unsigned)ceil(cycles/freq);    // Number of samples
    freq    = (double)cycles / samples;       // The actual frequency

    for (unsigned step = 0; step < samples; step++)
    {
      // Now step through the samples for this frequency, generating
      // the sine wave.
      double out = sin(phase) * amplitude;  // Calculate the raw number
      if (round > fabs(out * DBL_EPSILON))
      {
        // round off the answer and print.
        cout << round * floor(out / round + 0.5) << endl;
      }
      else cout << out << endl; // unless no rounding is indicated
      phase += 2 * M_PI * freq;             // update phase
    }
  }
}
```

Listing 9.3 The sine generation code (continued from previous page)

Listing 9.4 gives an example of the analysis code. Given the same set of parameters it generates the same frequency sequence as Listing 9.3, but in this case it reads in a line of system output at each sample step and performs demodulation described in (9.3) for each frequency step, then prints the result as text.

Of course, the code shown so far has a number of shortcomings that would need to be overcome in a real system: the Bode plot isn't actually printed, the format needed by Listing 9.4 is anonymous so the parameters must be remembered separately from the file, C++ isn't the best language for scientific computation of this sort, and so on. All of these issues can and should be overcome in a real system.

```
/************************************************************************
  name: analyzeSine

  description:  Analyzes a sine wave with the correct characteristics

  inputs: In addition to the parameters, expects input on cin in the
  form lines containing out1 and out2 seperated by whitespace.
  startF:     The start frequency (in fractional samples) of the sweep
  stopF:      The end frequency (in fractional samples) of the sweep
  minSamp:    The minimum number of samples per frequency step
  numStep:    The number of frequency steps

  outputs:
  none, but generates output on cout
************************************************************************/
static void analyzeSine(double startF, double stopF,
                        unsigned minSamp, unsigned numStep)
{
  double  logStartF = log(startF),
          logStopF  = log(stopF);

  double phase = 0.0;    // The output phase, so it can be kept continuous
  for (unsigned step = 0; step < numStep; step++)
  {
    // Step through the frequencies one by one until done
    double freq;
    unsigned cycles, samples;

    // Calculate the nominal frequency
    if (numStep < 2) freq = startF;
    else freq = exp(logStartF + step * (logStopF - logStartF) /
                                          (numStep - 1));
```

Listing 9.4 The output analysis code (continued on next page)

```
    // This block of statements calculates the constants necessary for the
    // output at this frequency to be an integer number of cycles with no
    // less than minSamp samples at the frequency.
    cycles  = (unsigned)ceil(freq * minSamp); // Number of cycles
    samples = (unsigned)ceil(cycles/freq);    // Number of samples
    freq    = (double)cycles / samples;       // The actual frequency

    complex<double> sum1(0.0, 0.0), sum2(0.0, 0.0);

    for (unsigned step = 0; step < samples; step++)
    {
      double out1, out2;
      char lineString[80];
      cin.getline(lineString, 79);
      stringstream line(lineString);
      line >> out1 >> out2;
      complex<double> K(exp(complex<double>(0.0, phase - M_PI/2)));

      sum1 += out1 * K;
      sum2 += out2 * K;
      phase += 2 * M_PI * freq;             // update phase
    }
    cout << freq << ", " << sum1 / (double)samples << ", ";
    cout << sum2 / (double)samples << endl;
  }
}
```

Listing 9.4 The output analysis code (continued from previous page)

9.6 Other Methods

This subject is a small subset of the overall subject of system identification and control. Other popular methods of designing control systems for plants with unknown characteristics are the Ziegler-Nichols and Åström-Hagglund methods for tuning PID controllers, ARMA system identification methods, and frequency response methods using random excitation. Each one of these methods has their advantages and adherents, and there are cases in control system design where one of these methods is clearly superior to the swept-frequency response measurement method shown here. I feel, however, that this method is the best overall method for tuning motion control loops in an embedded software environment. It has certainly served me well.

10

Software Implications

So far, we've discussed how to design control systems in the abstract: numbers, if they came into the discussion at all, were assumed to be real numbers with infinite precision and range. Real control systems implemented in real software do not have infinite precision numbering systems available to them. In the real world, numbers take space in memory to store. Perhaps more importantly, computations with them take time. Often, the biggest driver of processor size and cost in an embedded control system is the computational load imposed on the processor by the control algorithm. One's choice of numerical representation can have profound impact on the required processor speed, the amount of effort it takes to understand and implement the system in the first place, and the amount of effort it takes to maintain it through its life cycle.

10.1 Data Types

While there are probably as many different numerical representations as there are mathematicians, there are only a few that have proven useful in embedded control work. Floating-point representation and the various flavors of integers are the most familiar types, but the fixed-radix representation is also very worthwhile to know.

Integer

The binary representation for positive integers is just a simple base 2 conversion of the number, with the least significant bit of the number having a weight of 1 and the most significant bit of an N-bit number having a weight 2^{N-1}. Signed integer types can get more complex. For embedded processors, the ubiquitous signed integer type is two's complement. Other signed integer representations were popular in mainframes and minicomputers but never really made it into the microprocessor world. These formats may still be found if a controller is implemented in custom logic; they are signed-magnitude representation and one's complement. All of these number types share the characteristic that there is a sign bit that is true for negative numbers.

Twos complement is the most familiar signed integer representation. In twos complement the negative of a number is just 0 subtracted by the number, with any overflow ignored—so in four bit twos complement –5 is represented as 0 – 0101 = 1011. An N-bit twos complement number can represent any value from -2^{N-1} up to $2^{N-1}-1$. Note that with twos complement arithmetic, the negative range extends exactly one count farther than the positive range—if your software checks for overflow this can cause subtle bugs.

Ones complement represents a negative number as the bitwise complement of the number—so in four bit ones complement –5 is 1010. An N-bit ones complement number can represent any value from $-2^{N-1}+1$ up to $2^{N-1}-1$, with the possibility of having both "positive zero" (0000) and "negative zero" (1111).

Signed magnitude splits the number into a sign flag bit and uses the balance of the number as an unsigned magnitude. A four bit signed magnitude representation of –5 would be 1101. An N-bit signed magnitude number can represent any value from $-2^{N-1}+1$ up to $2^{N-1}-1$, and signed magnitude also allows both positive (0000) and negative (1000) zero.

Floating Point

Floating-point representation gives up some precision in the numerical representation, and uses the bits that are freed up to allow the number to represent a wider range. I won't try to describe any floating-point formats in detail here; I will only describe a few salient features of floating point. A typical floating-point representation will have an exponent (usually signed) and a mantissa (which is almost invariably signed). A typical 32-bit floating-point number would devote 8 bits to a signed exponent with the remaining 24 bits devoted to the mantissa; this typically allows a number with an approximate range of 10^{-37} to 10^{38}, with the ability to discern a difference of about 30 parts per billion between numbers.

Floating-point numbers are a great convenience in embedded control systems if you have the processing power available to you. Because the nature of a control system is such that you generally know (or can figure out) the range and reliability of every signal in the system, however, floating-point numbers are not a necessity. Many processors, even ones which do floating-point calculations in hardware, perform floating-point calculations significantly slower than integer calculations, and fast floating-point hardware can add significantly to the cost of a processor. In general floating-point calculations in embedded control systems should be considered a luxury unless the controller is a small part of the overall embedded system, or only very few systems are to be built.

Fixed Point

Computation with floating point is convenient but can be too expensive. Computation with integers is inexpensive, but it can be cumbersome in control systems where many of the multiplicands in computations are close to or less than 1. Fixed-point data types are a generalization of integer data types that overcome this difficulty by interpreting the bit-weighting of the data differently than an integer type.

In an integer data type each word has an implied radix point just to the right of the least significant bit, i.e., 0001 is interpreted as 1.0. Fixed-point types expand this interpretation and specify where the radix point is. I will use the notation IrQ, where I is the number of bits to the left of the radix point and Q is the number of bits to the right—so for example a 16-bit integer type would be 16r0. A purely fractional, signed, 16-bit number would have a least significant bit weight of 2^{-15}, with the sign bit having a weight of 1—therefore a signed fractional representation would be a 1r15 representation, with 0111 1111 1111 1111 representing nearly 1 (actually $1 - 2^{-15}$), and 1000 0000 0000 0000 representing exactly –1.

Many fixed-point DSP chips implement fractional or fixed-point arithmetic in their native machine code, and most "regular" processors have instructions that can be used to closely simulate fractional (1r15 or 1r31) arithmetic. This can significantly speed up computations if the code is written right. It is often worthwhile to write a library that implements fixed-point arithmetic in C or C++, even if all of the operations are synthesized in ANSI compatible code. Things get significantly better if you can write the actual computations in assembly language.

10.2 Quantization

One consequence of having a finite word size is that we can only represent quantities to a limited precision. An obvious example of this is the integer type, which can only represent numbers in steps of 1. As a result, for example, a controller using integers cannot distinguish between a computation, yielding a result of 25.3 and a computation yielding a result of exactly 25.

The effect of limited precision is to force us to choose a number from a finite number of quantities; in analysis this effect is called *quantization*. Figure 10.1 shows an example of quantization, in this case rounding. The ordinate axis is the desired or "naturally occurring" value, and the coordinate axis is the actual value that will result from quantization.

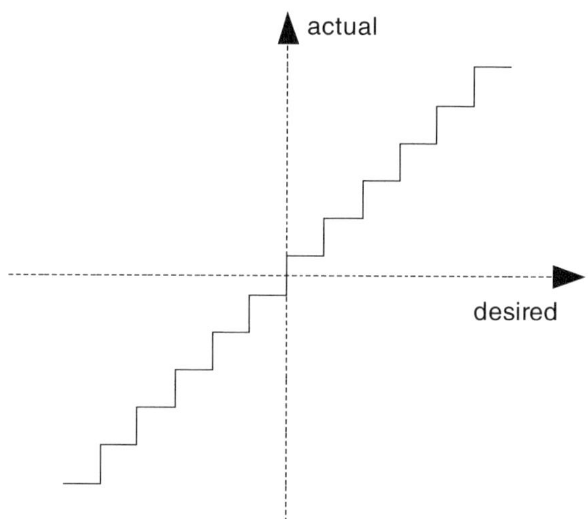

Figure 10.1 An example of quantization

Quantization happens in many places in a digital control system: When a value is measured, it gets quantized, each intermediate calculation is quantized, any numbers that are stored get quantized, and finally the ultimate controller output must be quantized before being applied to the outside world.

In general, you want to design your control system so that the only significant quantization that happens occurs when the input value is measured and when the control value is sent out. Unless your processor resources are severely constrained, it is fairly straightforward to ensure that this is the case, but it doesn't happen automatically: you must pay attention to your design to ensure that this happens.

Quantization Noise

Full theoretical treatments of how quantization affects a system in closed loop can be prohibitively difficult. You can simulate your system to see exactly what quantization will do, and indeed that is a good thing to do if you have an accurate system model in a form where it can be simulated. Simulations are sometimes difficult to construct and modify, however. Indeed, if you are in the early phases of a control system design where the control system architecture is not yet finalized it can be very time consuming to construct simulatable candidate system models just to assess the effect of quantization.

The solution to these problems is to treat quantization effects as noise. To model quantization as noise, rather than treating the quantizer output as a difficult to find and ambiguous, but deterministic, function of the input, you treat the quantizer output as the input plus some noise signal. Knowing the extent of the quantization you can make some assumptions about the nature of its impact and make predictions about its effect on your system's performance. In a system block diagram, this model of quantization replaces each quantization block with a simple gain, a summing junction, and injected noise, the level of which is the same as the level of quantization error expected.

Once you have decided to treat the quantization effects as noise you must choose a frequency spectrum for your noise model. In general you will either want to model the noise as purely random (that is, white noise that is spread evenly over the frequency spectrum), or as being periodic, and occurring at the very worst frequency that it possibly can. Generally, if you can be sure that the input to the quantized block is already fairly random, then it is safe to treat the quantization noise as added random noise. If, however, the prevailing noise level in the system will be less than the quantization noise it is best to treat the noise as occurring at the worst possible frequency.

To find the effect of quantization on a particular portion of your system, such as a filter, you need to model the subsystem as a block diagram, calling out each place where quantization will cause problems with an explicit quantization block. Replace the quantization block with a noise source, and do your analysis.

Figure 10.2 is a block diagram of a direct-form I filter with a noise input to model quantization noise (code to implement this filter in floating-point format is shown in Listing 10.1). With the noise inputs set to zero his filter implements the transfer function

$$\frac{Y}{U} = \frac{b_2 z^2 + b_1 z + b_0}{z^2 + a_1 z + a_0}, \tag{10.1}$$

and with the signal input set to zero it implements the transfer function

$$\frac{Y}{U} = \frac{b_2 z^2 + b_1 z + b_0}{z^2 + a_1 z + a_0}. \tag{10.2}$$

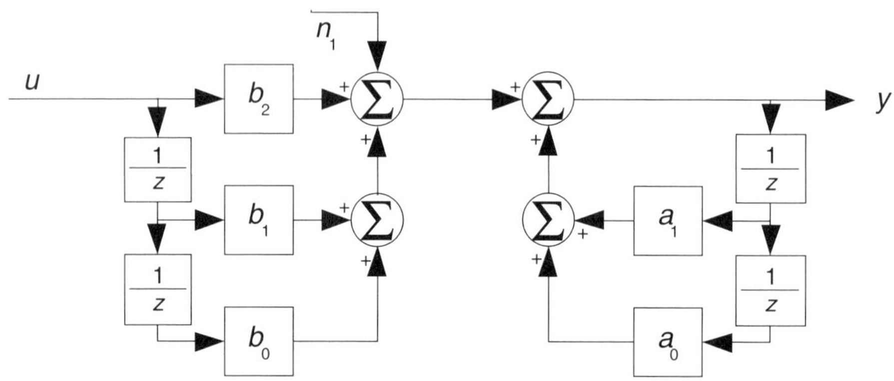

Figure 10.2 A direct-form I filter

For example, take the generic filter given in Figure 10.2, configured to implement a notch filter with a transfer function of

$$\frac{Y}{U}=\frac{0.95125z^2-1.88347z+0.95125}{z^2-1.88347z+0.9025}. \tag{10.3}$$

This transfer function has a response that is equal to 1 at $z = 1$ and $z = -1$, goes to zero at $z = 0.142$ ($0.142/T_S$ radians per second, or $0.0226/T_S$ Hz) and has a 3dB bandwidth of about $0.1/T_S$ radians/sec. Now assume that the input signal u is the output of a 12-bit ADC and ranges from –2048 to 2047. Further assume that you wish to hold disturbances at the notch frequency to half an LSB of the ADC.

Each multiplication block in Figure 10.2 adds quantization noise. Since there are five multiplication blocks, this provides five opportunities for quantization noise. At best this noise is uncorrelated, so the total noise power is five times the quantization noise of any one block. At worst the noise is correlated, so the total noise *amplitude* is five times the quantization noise of any one block. Each block's quantization noise will average out to ¼, so the noise contribution has an RMS amplitude of

$$\sqrt{\frac{5}{4}}q \le n_{RMS} \le \frac{5}{4}q, \tag{10.4}$$

where q is the quantization level.

Figure 10.3 shows the magnitude response of the filter to the intended signal and to the quantization noise. The intended response never goes above 1, but the response to noise peaks at around 77 times the quantization noise, and has noise bandwidth of $0.22/T_S$ radians/sec. At worst, all of the noise is concentrated at the frequency of the peak, which leads to an RMS output noise of up to $(77)(1.25q) = 96q$. At best, the noise is spread uniformly over the band, and the RMS output noise is $(0.22)(52.5)(1.12q) = 13q$.

In order to hold the noise from the notch filter to below ½ the value of q needs to follow

$$\frac{1}{192} \leq q \leq \frac{1}{26}, \tag{10.5}$$

which implies that the word size for the data must be at least 17 bits, and possibly 20.

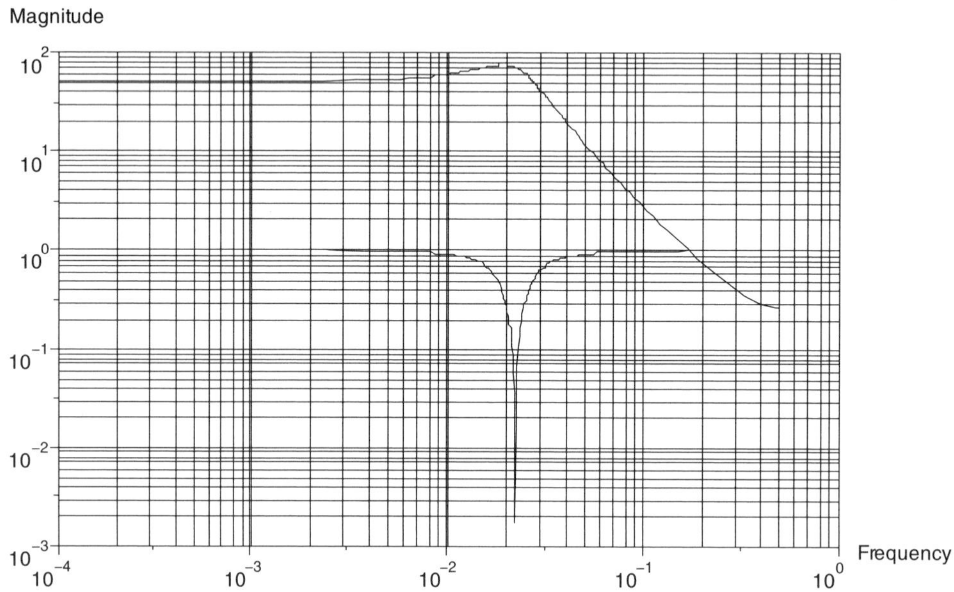

Figure 10.3 Frequency response of filter to signal and noise

```
// name: updateFilter
//
// description:
// A direct form I 2nd-order filter with the transfer function
//
//         num[2]*z^2 + num[1]*z + num[0]
//  H(z) = -----------------------------
//              z^2 + den[1]*z + den[0]
//
// inputs:
//   input:        Filter Input
//   numerator:    Pointer to an array of floats with 3 elements
//                 describing the transfer function numerator
//                 coefficients
//   denominator: Pointer to an array of floats with 2 elements
//                 describing the transfer function denominator
//                 coefficients
//   states:       Pointer to an array of floats with 2 elements
//                 containing the filter states.
//   oldInput:     Pointer to an array of floats with 2 elements
//                 containing the filter inputs
//
// returns:        The current filter output
////////////////////////////////////////////////////////////////
float updateFilter(float input, const float * num, const float * den,
                   float * states, float * oldInputs)
{
  int n;
  const unsigned int size = 2;
  float output = num[size] * input;

  // calculate the new output
  for (n = 0; n < size; n++)
  {
    output += num[n] * oldInputs[n] - den[n] * states[n];
  }

  // shift the inputs & outputs into their places
  for (n = 0; n < size - 1; n++)
  {
    oldInputs[n] = oldInputs[n+1];
    states[n]    = states[n+1];
  }

  states[size-1] = output;
  oldInputs[size-1] = input;

  return output;
}
```

Listing 10.1 A direct-form filter

Quantization Gain Effects

Another effect of quantization is to create a situation where the effective gain of the system varies between zero and infinity. This happens when the noise of the system is insignificant compared to the quantization level. Look at Figure 10.1 again and assume that you have a system that is fairly noise free, and that must settle to zero. Further assume that the system has a controller with some integral action, as might occur with the system in Figure 10.4. In this system the feedback is rounded to the nearest integer value, and the controller is a PI driving an integrating plant. With a target value set to 0.75, the behavior of the system is shown in Figure 10.5. The system output value rises to 1.0, then bounces around that value as the feedback jumps from +0.5 to 0.

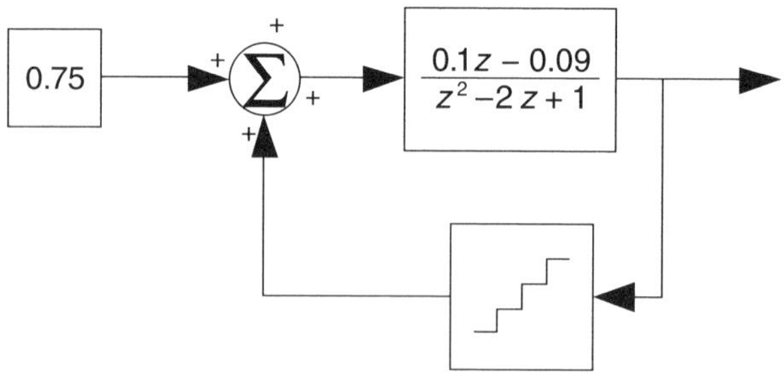

Figure 10.4 A system with quantization

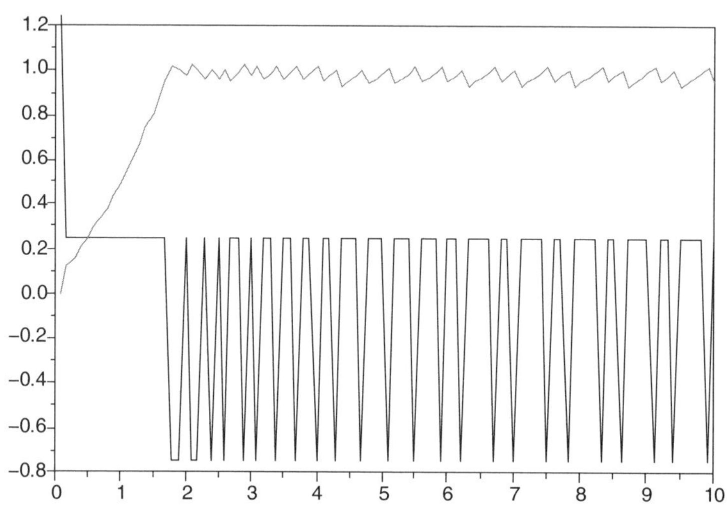

Figure 10.5 Response of a system with quantization

There are two elements of note to the system response shown in Figure 10.5. First, the system doesn't ever really settle—it oscillates around a steady value, but it never stops moving. This behavior is a direct result of the infinite gain of the quantization as the output switches values. Second, the system doesn't settle to the correct value. Because the desired final value is in between quantization steps the system cannot go there, so it hovers around 1.0, which is an incorrect value.

Quantization gain effects can cause the sort of oscillation seen in Figure 10.5, and it can cause the system to settle out at the incorrect value as seen in Figure 10.5. In some systems it can cause the system to drift around the correct value, without ever actually settling out.

Underflow

Perhaps the most insidious quantization effect for the beginning writer of control software is underflow. Underflow is nothing more than quantization gain effects as mentioned in the previous section, but the places where it occurs, the severity of its effects and the steps you should take to eliminate it are sufficiently different that it deserves its own discussion.

Underflow occurs when your code contains an integrator whose input magnitude is too small to affect its value. When this happens the system will not work correctly. Because integrator gains are often the smallest gains in a control system, one must pay close attention to underflow issues around one's integrators.

Take, for example, the simple low-pass filter described by

$$y_k = dy_{k-1} + (1-d)u_k \,. \qquad (10.6)$$

Nominally this has the transfer function

$$\frac{Y}{U} = \frac{z}{z-d}, \qquad (10.7)$$

which is an accurate description of the behavior of the filter *only* as long as the filter is operating in its linear region. Say that you've chosen a system design where the input is a 12-bit ADC that is right-justified (i.e., the least significant bit of the ADC has a weight of 1). If you choose a value of $d = 0.95$ and copy (10.6) into your code using integer arithmetic you may end up with something like Listing 10.2.

```
int dumbLowPass(int u) {
  static int y;
  y = (19 * y + u)/20;
  return y;
}
```

Listing 10.2 A quick low-pass filter

The code in Listing 10.2 will exhibit severe underflow. The division by 20 will quantize the result of the summation inside the parentheses to the point where the input variable will have to change by 20 steps before it can have an effect. This loss of precision will effectively reduce the precision of the 12-bit input value to a precision of less than 8 bits.

The solution when using integer arithmetic is to maintain the state variable y at a higher precision, and divide its value down to the correct return value at the end of the function, as shown in Listing 10.3. Here the value of the state variable is still quantized, but its quantization level exactly matches the quantization level of the input. In addition, the intermediate result is rounded rather than truncated (by adding half the divisor to the state before division). This filter will do a much better job than that of Listing 10.2.

```
int betterLowPass(int u) {
  static int state;
  state = state + u - (state + 10) / 20;
  return state / 20;
}
```

Listing 10.3 Scaling the state variable

We can take this one step farther by scaling up the input slightly, as shown in Listing 10.4. Here the input is multiplied by 5, and the return scaled back appropriately.

```
int evenBetterLowPass(int u) {
  static int state;
  state = state + 5 * u - (state + 10) / 20;
  return state / 100;
}
```

Listing 10.4 Scaling the input and state variables

Quantization with Floating Point

Quantization is fairly easy to understand when you are working with a fixed-point data type. It can be a bit more slippery when you use floating-point data, but you *can* get a handle on it. The quantization that occurs with floating point is not a fixed amount; rather, it depends on the magnitude of the numbers involved and the operation that is being carried out.

When two floating-point numbers are multiplied, the quantization is usually not a big issue, because floating-point multiplication uses the whole mantissas of both numbers. Quantization becomes an issue when you add or subtract numbers, however. Whenever two floating-point numbers have different exponents, the mantissa of the lower magnitude number is shifted down before the summation, which inevitably causes a loss of precision.

If floating point is used to implement an integrator, even the inherent integrators in filters, quantization will be an issue when a small input is added to a large integrator state. For example IEEE floating point specifies a 24-bit mantissa with an implied leading '1', so the accuracy of the data is limited to 25 bits relative to the state. This would be a problem, for instance, if you have a system that is being sampled very rapidly and so has a very small integrator gain.

Consider a system which uses floating point for its processing. It requires 16-bit input accuracy, and it normalizes all quantities to the range –1 to 1. Thus, the smallest increment that the input will see is 2^{-15}, or approximately 3×10^{-5}. With a floating-point value of 1.0, the smallest increment that can be tracked with a 25-bit mantissa is on the order of 3×10^{-8}—so if the integrator gain is less than 0.001, small values of the input will not affect the integrator value, and the effective input accuracy will be reduced.

Most systems that provide floating-point arithmetic provide 64-bit floating-point numbers, which have 48 bit mantissas. While a system with 16-bit input data might easily require more precision than is available from 32-bit floating-point numbers, it is a rare system that will tax the precision of a 64-bit floating-point number!

Filter Order

It is a general rule in DSP that a filter[1] should be implemented in a number of small parts, rather than lumping the whole filter together as one. Lumping the filter together will significantly increase its sensitivity to quantization noise and to coefficient quantization.

As an example, consider a controller that consists of a PID controller preceded by a low-pass filter. Its transfer function can be expressed in an infinite number of ways. In this case, I'll choose one that reflects the structure of the system:

$$H_c(z) = \left(100\frac{0.2(z-1)}{z-0.8} + 20 + \frac{0.1}{z-1}\right)\frac{0.8z}{z-0.8}, \tag{10.8}$$

and one that reduces the transfer function as much as possible:

$$H_c(z) = \frac{14z^3 - 27.196z^2 + 13.1968z}{z^3 - 2.6z^2 + 2.24z - 0.64}. \tag{10.9}$$

[1] At least an infinite impulse response filter, which is the kind we deal with here.

One could take the direct-form filter implemented in Listing 10.1, extend it to a third-order filter, and use it to implement the transfer function in (10.9) directly. If this were done, however, a number of problems would arise. First, any quantization of the coefficients of (10.9) will strongly affect its properties, well out of proportion to the quantization itself.

For example, allow the z^0 coefficient to change by 0.00003 (roughly 2^{-15}), so (10.9) becomes

$$H_c(z) = \frac{14z^3 - 27.196z^2 + 13.1968z}{z^3 - 2.6z^2 + 2.24z - 0.63997}. \tag{10.10}$$

This seemingly insignificant change in one coefficient makes a drastic change in the magnitude response of the system, as shown in Figure 10.6: instead of an integrator with infinite gain at DC, the system now has a low-pass filter with a DC gain that's barely 6dB above the proportional gain of the system. Using this form of the controller would have consequences on the DC accuracy of the system ranging from mild to serious.

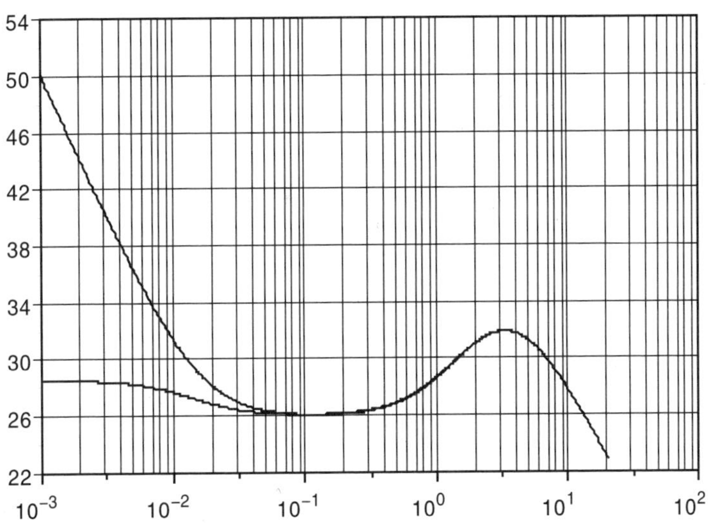

Figure 10.6 Magnitude plot of two realizations of $H_c(z)$

This sort of quantization impacts filters in rough proportion to two factors: the closeness of the filter's poles and zeros to 1.0, and the order of the filter. The way to avoid this problem is always to break your controller into first-order filters where possible. If necessary (in the case of a notch filter or other filter with complex poles) you will have to use second-order filters, but you should never have to go higher than a second-order filter.

In addition to coefficient quantization, filters also get more sensitive to quantization in the states as the filter order goes up. The rule of thumb is that for a filter with poles at $z = d$ you need enough precision to store your smallest input value times $(1 - |d|)^n$, where n is the filter order. This can become a challenge with second-order filters, and should just not be contemplated with higher order filters.

10.3 Overflow

As we go from the filter in Listing 10.2 to that of Listing 10.3 and then Listing 10.4 we do an increasingly good job of solving the underflow problem—but we risk running into an issue with word size. The input is stated to have a 12-bit range. Assuming that's –2048 to 2047, the filter in Listing 10.3 can have a state value that ranges from –40960 to 40940, and the Listing 10.4 filter can have the state value range from –204800 to 204700. Clearly, neither of these latter two filters will work if the state values are stored in 16-bit integers!

This overflow is the flip side to underflow and quantization noise. Quantization causes problems when your numerical representation doesn't have sufficient accuracy to resolve the numbers that need to be resolved. Overflow isn't a problem with small numbers, but it does become a problem when a result of a computation is too big to fit into the available space. You should ensure that your computations will not overflow unexpectedly, and if possible you should insure that any overflows that do happen are as benign as possible.

Rollover

Most microprocessors implement their arithmetic such that integers overflow modulo 2^N on any addition, subtraction or multiply for an N bit number. For a computation whose result using infinite precision would be x_d the actual result would be

$$x_a = x_d mod 2^N , \tag{10.11}$$

For example, the sum of 5 + 7 is 12. On a four-bit processor this would be 1100—but if it were a twos complement number, the value of the result would be –6, not 12! This rollover effect happens with any size number, be it 16- or 32-bit. Worse, if you're programming in C or C++, there are no run-time checks for overflow unless you add them yourself. Of course, in a control system, getting a large positive control output when you needed a large negative output is about the worst thing that could happen.

Overflow with Floating Point

Overflow is generally something that happens to integers, but not to floating-point numbers (unless you have a very odd scaling of your variables!). What *can* happen, however, is for a variable to grow to the point where the required precision is not met, and all reasonable computations underflow. Once this happens, your filters will not work until their states are reset and everything starts out over again.

Saturation

Possibly the most benign form of overflow that can happen in a control system is saturation. In saturation the overflow takes the form

$$x_a = \begin{cases} x_{min} & x_a \le x_{min} \\ x_a & x_{min} < x_a < x_{max} \\ x_{max} & x_a \ge x_{max} \end{cases} , \tag{10.12}$$

where x_{min} is the minimum possible value that can be represented and x_{max} is the maximum possible value. In other words, if the result of a computation violates the bounds of what can be represented in the numbering system, the processor recognizes this fact and comes as close as it can to the correct value.

Saturation in this form is usually very benign. In fact, it is routine to saturate values intentionally (in integrators, and before they are applied to output channels, for instance). Whenever a value is to be applied to a variable or an output with a lower range, its value should be checked and saturated rather than allowing the value to roll over or otherwise get mangled.

10.4 Resource Issues

Computation time

Controllers require a lot of computations, and computations can soak up an amazing amount of processing power. The amount of time that a processor spends in updating a controller and computing its output can easily become the most significant performance driver in an embedded control system.

In general, the amount of time that your processor spends on computation is important for three reasons: one, because it directly affects processor loading; two, because it creates a lag between the input sampling time and the time the DAC can be loaded, and finally because if the computation time should vary, it can cause the output sampling times to jitter.

Early in a project, it can be hard to pin down the amount of time a control computation will take. Different processors have significantly different timing characteristics when performing computations, and there aren't any good general rules for predicting how fast a processor may compute a particular control rule. The way that a controller is coded can have a significant impact on the speed of computations, and there is often a distinct tradeoff between code readability and execution speed.

The only sure fire way to predict how much time a control loop will take to execute on a given processor is to write some prototype code and try it out. Sometimes this can be done on a cycle-accurate simulator if one takes I/O and memory speed into account; sometimes the only way to go is to get an evaluation board and give the code a whirl.

All is not lost, though: with experience and a well-annotated instruction set reference you can make a good estimate. I must stress the level of experience required (and point out that I've made some astounding misses in this regard). I don't recommend this approach until you've calibrated yourself by trying it out on a few processors that you're already familiar with, and can prototype control code on.

Computation time on any given processor depends heavily on a number of factors: how much raw speed does the processor have, what hardware does it have to speed up computation, what is its word length, and how much readability and/or code flexibility are you willing to give up in the search for fast execution on an inexpensive processor.

8-bit processors often lack hardware multipliers; when hardware multipliers exist, they will likely be 8x8 and must be extended for more precision. Because these chips can be very cost effective, particularly in high-volume applications, it can be worthwhile to consider them as candidates for lower-speed applications where you can afford to program them in assembly language.

If you are programming an 8-bit processor in assembly language, you can often gain significant amounts of performance by using shift-and-add multiply routines with reduced-precision coefficients. Many control loops can be implemented with no more than three significant bits in the coefficients. You can gain significant speed advantages by using custom 8-bit floating-point numbers with 3-bit mantissas, and you can gain even more if you hard-code your multiplications in assembly to reduce the number of operations to a minimum.[2]

A final trick to play with an 8-bit processor is to tailor the size of each part of the data path. Because you often need more precision in your integrators than anywhere else you can often get by with 16-bit states in your differentiators and 16-bit data paths elsewhere in your controller, but with a 24-bit integrator.

2 With a good macro assembler you can do this with a parameterized macro for readable, fast code—just don't ask anyone to understand the macro.

A 16-bit processor is certainly going to be faster than a comparable 8-bit one, and a 32-bit processor will be even more so. If it includes a reasonable hardware multiply and 16-bit data paths to memory, a 16-bit processor can perform many more computations per clock tick than an 8-bit processor. With a 16-bit processor, one is less likely to need to resort to significant amounts of assembly language, and any assembly that is used to speed up computations can be decently hidden in a library.

Don't overlook DSP chips when you are contemplating a controller. Designing controllers is in many ways just a specialized form of digital signal processing, so DSP chips have many features that will speed up controller computations. These processors have plenty of potential, but their most profound performance gains can only be achieved by vectorizing the controller computations and solving the resulting matrix multiply using the chip's multiply-and-accumulate instruction. Even without such measures, however, many DSP chips possess enough raw speed to make the controller performance issue an easy one to deal with.

Order of Computation

In linear system analysis the order of computation makes no difference to the final outcome: by virtue of the property of superposition the ultimate signal coming out of a system will be the same no matter the order of the intermediate steps.

However, quantization and overflow effects are not linear effects, so the order in which computations are done can have a strong effect on the system behavior. Moreover, some computational arrangements can be significantly more efficient, or make it much easier to write readable code, than others.

One example is an integrator. Take an integrator, the transfer function of which is

$$H(z) = \frac{k_i z}{z-1}. \tag{10.13}$$

This can either be implemented as a multiply followed by a summation, or as a summation followed by a multiply. If we were to implement the integrator with integer data types, with a right shift to implement the gain (integrator gain is almost always small), we may have code like in Listing 10.5. In the code on the left we will suffer severe quantization problems unless the input data is intentionally pre-shifted up to larger values; the code on the right will not suffer from quantization. Note that neither of these pieces of code have any protection against overflow.

```
int i_state;
...
i_state += input >> INT_SHIFT;
return i_state;
```

```
int i_state;
  ...
i_state += input;
return i_state >> INT_SHIFT
```

Listing 10.5 Two ways of implementing integrators with integers

Another example is a differentiator. Take a differentiator, the transfer function of which is

$$H(z) = k_d \frac{1-d}{z-d}. \tag{10.14}$$

Like the integrator, this can be implemented as a multiply followed by a difference, or as a difference followed by a multiply, as shown in Listing 10.6. Implementing the gain first, such as the case on the right, significantly increases the chances of overflow, since differential gain is almost always large. The computation on the left reduces the chances of overflow without losing any precision.

```
int d_state;
...
retval = input - d_state;
d_state = input;
return retval * d_gain;
```

```
int d_state;
...
input = d_gain * input;
retval = input - d_state;
d_state = input;
  return retval
```

Listing 10.6 Two ways of implementing integrators with integers

10.5 Implementation Examples

A Fractional Data Type

As pointed out earlier, using floating-point arithmetic is often impractical for control systems design because of issues with library robustness or execution times. The obvious answer to this is to use integer arithmetic for implementing the control system. This is sometimes a good answer, because integer arithmetic is available on a wide variety of platforms and it is well understood. Once you take the need to avoid problems with under- and overflow into account, however, integer arithmetic can become very cumbersome very quickly.

A solution to this problem is to use some code that implements a fractional data type that also reduces the problems with overflow. Such a fractional data type should at least offer a representation with numbers that range from –1 to 1 and which, for any operation that may result in overflow, will saturate the result rather than allowing it to wrap.

Listing 10.7 shows a C header file for a fractional data type that meets the above requirements. The file defines a number of functions that, when used correctly, will streamline the design of control systems. The functions can be split into two varieties: the first is to enable addition and multiplication by integers in a way that will saturate on overflow instead of rolling over; while the second will perform multiplies and divides in a fractional number space, where an integer is redefined as ranging from –1 to 1.

```
#define   FR_SHIFT    (sizeof(fr_type) * CHAR_BIT - 1)
#define   FLOAT_DIV   ((double)INT_MAX + 1.0)

typedef int fr_type;

fr_type sat_add(fr_type a, fr_type b);    // Add and saturate

fr_type sat_sub(fr_type a, fr_type b);    // Subtract and saturate

fr_type sat_mul(fr_type a, fr_type b);    // Multiply and saturate

fr_type fr_mul(fr_type a, fr_type b);     // Multiply by fractional value

fr_type div_to_fr(fr_type a, fr_type b);  // Divide to fractional value

fr_type div_to_int(fr_type a, fr_type b); // Divide to integer value

fr_type make_fr(double a);                // make from float

double get_fr(fr_type a);                 // extract float
```

Listing 10.7 Header file for a fractional data type

This data type can be implemented in one of three ways. The most direct method is to choose a storage type for the data type that is not the largest available from the compiler and to leverage the larger data type for multiplications. Doing so will result in code that is fairly portable, moderately fast, and very easy to understand. The second method is to implement the functions in assembly language. On most machines this has the advantage of accessing overflow detection operations that are not available from most high-level languages, yielding faster, more compact code. The third method is to use the largest data type that is available on the compiler and to synthesize the multiplications. This has the advantage of being portable across a wide range of platforms, but the code can be extremely cryptic. The first two methods are presented here, using GNU C (and assembly) on the PC.

It is handy to have one very portable copy of this code for use on any processor. This will serve for applications that do not need to wring the last bit of speed from a processor, and for the initial development in applications that will eventually need the speed. For applications that *do* need all the processor speed available, it is wise to replace all of the C functions with the corresponding functions in assembly as the speedup is generally a factor of 2 to 10.

Do not assume, however, that this is the best approach for every processor. There are a few processors out there—including newer Pentiums and floating-point DSP chips—which will perform floating-point calculations faster than they will these fixed-point ones. Also, there are many processors that are nearly as fast using floating point as they are with this fixed-precision type, and there are some control problems that simply don't demand that much from a processor. Certainly, if you have the space and the speed to use floating-point math, this code may not be the most cost-effective.

Saturating Addition and Subtraction

The first three functions implement arithmetic with saturation: the normal operation is performed, but on overflow the result is saturated at the top of the range. Listing 10.8 and Listing 10.9 show the C code for saturated addition and saturated subtraction respectively. Both of these functions assume that they are dealing with twos complement math which rolls over on overflow; they detect overflow by observing an incorrect sign in the result.

```
#include "fr_type.h"
#include <limits.h>

fr_type sat_add(fr_type a, fr_type b)
{
  fr_type retval = a + b;    // Do a regular add

  // Check for rollover and saturate if so
  if (a < 0 && b < 0)
  {
    if (retval > 0 || retval < -INT_MAX) retval = -INT_MAX;
  }
  else if (a > 0 && b > 0 && retval < 0) retval = INT_MAX;

  return retval;
}
```

Listing 10.8 Saturating addition in C

```
#include "fr_type.h"
#include <limits.h>

fr_type sat_sub(fr_type a, fr_type b)
{
  fr_type retval = a - b;    // Do a regular subtraction

  // Check for rollover and saturate if so
  if (a < 0 && b > 0)
  {
    if (retval > 0 || retval < -INT_MAX) retval = -INT_MAX;
  }
  else if (a > 0 && b < 0 && retval < 0) retval = INT_MAX;

  return retval;
}
```

Listing 10.9 Saturating subtraction in C

Listings 10.10 and 10.11 show the same functionality implemented in assembly. Here the overflow flag is used as intended, with an additional test for an operation that results in 0x80000000—this is done because in twos complement notation this value, when negated, equals itself; keeping it out of the numbering system has great advantages in preventing a variable from "sticking" at the negative extreme!

```
.file "sat_add.x86.s"
.globl _sat_add
    .def     _sat_add;     .scl     2;     .type       32;     .endef
_sat_add:
        movl  4(%esp), %eax

        add   8(%esp), %eax        # do the add
        jno   k0                   # test for overflow
        jg    k1                   # it's overflow, check sign
        movl  $0x80000001, %eax    # saturate negative
        jmp   k0
k1:     movl  $0x7fffffff, %eax    # saturate positive
k0:     cmp   $0x80000000, %eax    # count -(INT_MAX+1) as overflow
        jne   k2
        movl  $0x80000001, %eax    # saturate negative

k2:     ret
```

Listing 10.10 Saturating addition in assembly

```
.file "sat_sub.x86.s"
.globl _sat_sub
    .def    _sat_sub;    .scl    2;    .type     32;    .endef
_sat_sub:
        movl       4(%esp), %eax

        sub   8(%esp), %eax        # do the subtraction
        jno   k0                   # test for overflow
        jg    k1                   # it's overflow, check sign
        movl  $0x80000001, %eax    # saturate negative
        jmp   k0
k1:     movl  $0x7fffffff, %eax    # saturate positive
k0:     cmp   $0x80000000, %eax    # count -(INT_MAX+1) as overflow
        jne   k2
        movl  $0x80000001, %eax    # saturate negative

k2:     ret
```

Listing 10.11 Saturating subtraction in assembly

Saturating Multiplication

Listing 10.12 shows a saturated multiplication routine in C. Unlike addition, you cannot check for overflow by looking for a sign change, because there are possible overflow combinations that will preserve the correct sign.

```
#include "fr_type.h"
#include <limits.h>

fr_type sat_mul(fr_type a, fr_type b)
{
  // Do a multiply into something guaranteed to contain the result
  long long result = (long long)a * (long long)b;

  // Check for rollover and saturate if so
  if (result > INT_MAX) result = INT_MAX;
  if (result < -INT_MAX) result = -INT_MAX;

  return result;
}
```

Listing 10.12 Saturating multiplication in C

Listing 10.13 shows saturating multiplication in assembly. Here again the overflow flag is used to detect overflow, and the "special" value 0x80000000 is not allowed to creep into the numbering system. This routine leverages the x86 trait of doing an N by N multiply and giving back a 2N result; if your processor has it then you can do something similar, otherwise you'll have to find a different way to test for overflow so that you can saturate when appropriate.

```
.file "sat_mul.x86.s"
.globl _sat_mul
   .def     _sat_mul;     .scl     2;      .type 32;     .endef
_sat_mul:
       movl    4(%esp), %eax

       imull   8(%esp)         # do the multiply
       jo      k1              # test for overflow
       cmp     $0x80000000, %eax # test for eax == -INT_MAX,
                                   #overflow if so
       jne     k0              # no overflow

k1:    cmp     $0, %edx        # there is overflow, so
       jl      k2              # test sign and saturate
       movl    $0x7fffffff, %eax    # saturate positive
       jmp     k0
k2:    movl    $0x80000001, %eax # saturate negative

k0:       ret
```

Listing 10.13 Saturating multiplication in assembly

Fractional Multiplication

This operation is the one that defines the data type. Fractional multiplication is performed such that the effective scaling of the data is 2^{-N+1}; in other words, for 32-bit data the least significant bit has a weight of 2^{-31}. In C this done simply with a multiply to a larger data type and a shift down by N–1, as shown in Listing 10.14. It is not necessary to test for overflow with fractional multiplication, because the result is guaranteed to have a smaller magnitude than either operand.

```
#include "fr_type.h"
#include <limits.h>

fr_type fr_mul(fr_type a, fr_type b)
{
  long long result = (long long)a * (long long)b;
  return result >> FR_SHIFT;
}
```

Listing 10.14 Fractional multiplication in C

The fractional multiply is ever so slightly strange in the way the sign bit behaves; because of this, the multiply to a larger data type works fine (here the processor multiplies the numbers to the edx:eax pair), but the result is now off by a factor of two. This is taken care of with a shift, and the result is returned, as shown in Listing 10.15.

```
.file "fr_mul.x86.s"
.globl _fr_mul
    .def      _fr_mul;   .scl   2;      .type     32;     .endef
_fr_mul:
        movl    4(%esp), %eax

        imull  8(%esp)          # do the multiply
        rol    %eax             # shift up one, save the carry bit
        rcl    %edx             # shift up one, get the carry from last
        mov    %edx, %eax

        ret
```

Listing 10.15 Fractional multiplication in assembly

Division

There are two types of division supported. One divides two integers (or fractional numbers) to an integer, saturating on overflow. The other divides two integers (or fractional numbers) to a fractional. Both of these divide routines return zero if a division by zero is attempted.[3]

Listing 10.16 shows the divide to integer routine, while Listing 10.17 shows the divide to fractional.

```
#include "fr_type.h"

int div_to_int(fr_type a, fr_type b)
{
  if (b == 0) return 0;        // Stop gracefully if illegal divide
  return sat_add(a, b/2) / b; // Round, divide, and return.
}
```

Listing 10.16 Divide to integer in C

```
#include "fr_type.h"
#include <limits.h>
fr_type div_to_fr(int a, int b)
{
  if (b == 0) return 0;    // Bomb if illegal divide

  // Shift a up to simulate a fractional value
  long long result = (long long)a << FR_SHIFT;
  result /= (long long)b;

  // Check for rollover and saturate if so
  if (result > INT_MAX) result = INT_MAX;
  if (result < -INT_MAX) result = -INT_MAX;

  return result;
}
```

Listing 10.17 Divide to fractional in C

---

[3] This is to avoid the processor locking up from a divide by zero exception. If you use assertions for debugging it may be a good idea to assert on a divide-by-zero rather than quietly returning zero.

Conversion

Of course, if you are using a platform that supports floating-point operations it is convenient to use floating point for the occasional calculation, even if it is not fast enough to use in a control loop. Listing 10.18 and Listing 10.19 show the routines to convert from floating point to fractional and back.

```
#include "fr_type.h"
#include <limits.h>

fr_type make_fr(double a)
{
  a = a * FLOAT_DIV + 0.5;
  a = floor(a);
  if (a > INT_MAX) return INT_MAX;
  else if (a < -INT_MAX) return -INT_MAX;

  return (fr_type)a;
}
```

Listing 10.18 Convert from floating point to fractional

```
#include "fr_type.h"
#include <limits.h>

double get_fr(fr_type a)
{
  return a / FLOAT_DIV;
}
```

Listing 10.19 Convert from fractional to floating point

Integrators

Integrators should be written to be readable, they should not have quantization problems to speak of, and they should have anti-windup properties (discussed in Chapter 6). An integrator algorithm that satisfies these requirements is

$$x_n = \begin{cases} x_{max} & x_{n-1} + k_i u_n > x_{max} \\ x_{n-1} + k_i u_n & x_{min} \le x_{n-1} + k_i u_n \le x_{max} \\ x_{min} & x_{n-1} + k_i u_n < x_{min} \end{cases}, \quad (10.15)$$

where x_{min} and x_{max} are the integrator's maximum and minimum allowable values, x is both the integrator state and its output, k_i is the integrator gain and u is the integrator input. As long as this integrator stays within its bounds, it can be linearized directly to the transfer function

$$H_i(z) = \frac{k_i z}{z-1}. \quad (10.16)$$

Listing 10.20 shows the C code to implement such an integrator using floating-point arithmetic. The code is quite direct, which is the advantage of floating point. This code has two features which will slow it down in practice: first, the fact that it uses floating point will, on many processors, cause a speed hit; second, the use of a data structure with indirect addressing will also slow it down to some degree, depending on the processor used.

```
typedef struct integrator_struct
{
  double _state;        // integrator state
  double _max, _min;    // integrator limits
  double _gain;         // integrator gain
} integrator_struct;

double integrator_update(integrator_struct * p, double input)
{
  p->_state += p->_gain * input;
  if (p->_state > p->_max) p->_state = p->_max;
  else if (p->_state < p->_min) p->_state = p->_min;
  return p->_state;
}
```

Listing 10.20 An integrator using floating point

On many processors the code can be sped up by using integer arithmetic, at the price of sacrificing some readability. Listing 10.21 shows code to implement an integrator with integer arithmetic. This code modifies the algorithm given in (10.15) by having the integrator gain operate on the integrator output instead of its input, but as written its transfer function is still represented by (10.16).

```
typedef struct integrator_struct
{
  int _state;          // integrator state
  int _max, _min;      // integrator limits
  int _gain,           // integer part of integrator gain
      _shift;          // right-shift to implement integrator gain
} integrator_struct;

int integrator_update(integrator_struct * p, int input)
{
  p->_state += input;
  if (p->_state > p->_max) p->_state = p->_max;
  else if (p->_state < p->_min) p->_state = p->_min;
  return ((p->_state) * p->_gain) >> p->_shift;
}
```

Listing 10.21 An integrator using integers

An integer multiplication can only increase a value, while integrator gains are nearly always less than one. This requires that the integrator gain be implemented as an integer-part gain and a shift (it could have been implemented as a divide at the cost of more processing time). To alleviate quantization problems, the integrator gain operates on the integrator state rather than the integrator input. This latter difference has little impact if the system is tuned and left alone, but during tuning any change in integrator gain will cause a jump in the integrator output, which is probably not desirable. At best this is an inconvenience; if the system requires gain scheduling, it would cause difficulties during operation.

With the method of gain implemented, it is important that the possibility of overflow be checked and eliminated. The two points where overflow is going to occur are either where the state is multiplied by the integer part of the gain, or where the state is incremented by the input but not yet limited. You must know the gain, the allowable limits, and the possible range of input values to verify that you will not experience overflow.

Because the integrator gain operates on the integrator state, the meaning of the integrator limits is different from the floating-point code in Listing 10.20. If the integrator limits need to hold the output at certain points as is nearly always the case these limits become dependent on the integrator gain, and must be changed as the integrator is tuned.

Listing 10.22 shows the integrator code from Listing 10.20, but this time using the fractional arithmetic code. The integrator gain is assumed to always be less than 1; other than that this function sports all of the same advantages of the floating-point function, with the additional advantage that it won't cause nearly as great a speed hit on a cost-effective processor as would floating-point computations.

```
typedef struct integrator_struct
{
  fr_type _state;         // integrator state
  fr_type _max, _min;     // integrator limits
  fr_type _gain;          // integrator gain (always less than 1)
} integrator_struct;

fr_type integrator_update(integrator_struct * p, fr_type input)
{
  p->_state = sat_add(p->_state, fr_mul(p->_gain, input));
  if (p->_state > p->_max) p->_state = p->_max;
  else if (p->_state < p->_min) p->_state = p->_min;
  return p->_state;
}
```

Listing 10.22 An integrator using fractional arithmetic

Low-Pass Filters

A first-order low-pass filter should be a simple thing to implement, and it is. There are a number of ways to do it, however, each with its own issues. Low-pass filters should have a predictable relationship between their input and their state, and it should be easy to verify that they won't have quantization problems and that they won't overflow. The most direct algorithm for a low-pass filter is

$$x_n = dx_{n-1} + u_n \tag{10.17}$$

where d is the filter pole, u is the input and x is the filter output and state. It is usually most convenient for a low-pass filter to have a DC gain of 1, which is not the case for the preceding filter. This situation can be corrected, while still only implementing one multiply, with the algorithm

$$x_n = x_{n-1} + (1-d)(u_n - x_{n-1}). \tag{10.18}$$

This implements a filter with the transfer function

$$H_{lp}(z) = \frac{(1-d)z}{z-d}. \tag{10.19}$$

Listing 10.23 shows the code for a low-pass filter implemented using floating point.

```
typedef struct lowpass_struct
{
  double _state;        // filter state
  double _gain;         // filter gain = 1 - pole
} lowpass_struct;

double lowpass_update(lowpass_struct * p, double input)
{
  p->_state += p->_gain * (input - p->_state);
  return p->_state;
}
```

Listing 10.23 A low-pass filter using floating point

As with integrators, integer arithmetic raises difficulties with low-pass filters. Listing 10.24 shows code for a low-pass filter. The same scheme is used with an integer gain and a shift, and much the same problems with overflow exist. The filter in Listing 10.23 will have a state that asymptotically approaches the average value of the input, and will never exceed the bounds of the input. This integer version of the filter, however, will have a state which can go as high as an input extreme shifted *up* by the shift value and divided by the gain. Moreover, the highest value (in the innermost parenthesis of the first line of the function) will be the input extreme shifted up by the shift value.

```
typedef struct lowpass_struct
{
  int _state;           // filter state
  int _gain,
      _shift;           // filter gain * 2^{-shift} = 1 - pole
} lowpass_struct;

double lowpass_update(lowpass_struct * p, double input)
{
  p->_state += input - ((p->_gain * p->_state) >> p->_shift);
  return (p->_gain * p->_state) >> p->_shift;
}
```

Listing 10.24 A low-pass filter using integer arithmetic

Once again the filter can often be implemented using fractional arithmetic rather than floating point, with gains in processor speed over floating point and convenience over integer math. Listing 10.25 shows such a filter. The filter gain can be counted on to be less than 1 for all stable filters so the filter state will not, by definition, overflow. Furthermore, as long as the input value is scaled to be as large as possible, the chance of underflow will be minimized (but not eliminated: if necessary, a larger integer type must be used as the underlying storage in fr_type).

```
typedef struct lowpass_struct
{
  int _state;       // filter state
  int _gain,
  _shift;           // filter gain * 2^{-shift} = 1 - pole
} lowpass_struct;

fr_type lowpass_update(lowpass_struct * p, fr_type input)
{
  p->_state = sat_add(p->_state,
              fr_mul(p->_gain, sat_sub(input, p->_state)));
  return p->_state;
}
```

Listing 10.25 A low-pass filter using fractional arithmetic

Differentiators

The most straightforward way of implementing a differentiator mirrors the development of the argument for a differentiator in a system: first we note that a differentiator must be included, hence we do so with something like

$$H_d(z) = k_d \frac{z-1}{z}. \tag{10.20}$$

Once this is done, we realize that we should flatten off the differentiator's high-frequency response, so we follow it by a low-pass filter like the one described in (10.19).

This works, but it is somewhat wasteful of processor cycles. Moreover, if the order of operations is to take the derivative, multiply by the deriviative gain, then low-pass filter, the chances of an overflow at the multiplication by derivative gain is high.

An alternative is to start with the band-limited differentiator transfer function,

$$H_d(z) = k_d \frac{(1-d_d)(z-1)}{z-d_d}, \tag{10.21}$$

and recognize that it can be developed from a low-pass filter:

$$1 - \frac{1-d_d}{z-d_d} = \frac{z-1}{z-d_d}. \tag{10.22}$$

With a bit of scaling, we get

$$H_d(z) = k_d(1-d_d)\left(1 - \frac{1-d_d}{z-d_d}\right). \tag{10.23}$$

While this is not the perfectly straightforward thing of beauty that we may like, it has some distinct advantages. First, the overflow characteristics of the state variable are the same as for the low-pass filter we just investigated—so if you are using integer math you know how to check for overflow, and if you're using fractional math you know the state *won't* overflow. Second, the underlying low-pass filter will operate on the input rather than its derivative so it will be much smoother. Third, it only uses one state, as opposed to the two required for a differentiator followed by a low-pass.

Listing 10.26 shows the code for the differentiator using floating point. I've included a load function which takes the desired differentiator gain and pole position as inputs and loads the correct values in to the structure.

```
typedef struct derivative_struct
{
  double _state;        // filter state
  double _pGain;        // filter gain = 1 - pole
  double _dGain;        // derivative gain * (1 - pole)
} derivative_struct;

void derivative_load(derivative_struct * p,
                     double pole, double dGain)
{
  p->_pGain = 1.0 - pole;
  p->_dGain = dGain * _pGain;
}

double derivative_update(derivative_struct * p, double input)
{
  double retval = p->_dGain * (input - p->_State);
  p->_state += p->_pGain * (input - p->_state);
  return retval;
}
```

Listing 10.26 A differentiator using floating point

Listing 10.27 shows a differentiator using integer arithmetic. The value of `temp` is a scaled version of the desired differential value; with some thought you can see that

$$x_{temp} = \frac{z-1}{z-1+2^{-pShift}k_{pGain}}. \tag{10.24}$$

Thus, the overall transfer function of this filter is

$$H_d(z) = k_{dGain}\frac{z-1}{z-1+2^{-pShift}k_{pGain}}. \tag{10.25}$$

Of course, it makes a great deal of sense to create a load function for this filter—even more so than the floating point. This load function is not included here for brevity, but it is on the CD-ROM, and can be written using Listing 10.26 and (10.25) as guides. Note that it is assumed that the differential gain will always be large enough that sufficient precision will be provided by an integer `_dGain`—if this is not the case a larger `_dGain` can be used with a right-shift to restore the correct magnitude.

```
typedef struct derivative_struct
{
  int _state;          // filter state
  int _pGain,          // filter gain * 2^{-shift} = 1 - pole
      _pShift;
  int _dGain;          // derivative gain * (1 - pole)
} derivative_struct;

int derivative_update(derivative_struct * p, int input)
{
  int temp;
  temp = input - ((p->_pGain * p->_state) >> p->_pShift);
  p->_state += temp;
  return p->_dGain * temp;
}
```

Listing 10.27 A differentiator using integer arithmetic

The differentiator using integer arithmetic suffers from all of the drawbacks of any of the other integer-based filters discussed so far. Listing 10.28 shows how the differentiator may be implemented using fractional arithmetic. The implementation is largely the same as the implementation of the floating-point differentiator. Once again, we are assuming that there will always be sufficient derivative gain to give `_dGain` a satisfactory precision.

```
typedef struct derivative_struct
{
  fr_type _state;          // filter state
  fr_type _pGain;          // filter gain = 1 - pole
  int     _dGain;          // derivative gain * (1 - pole)
} derivative_struct;

fr_type derivative_update(derivative_struct * p, fr_type input)
{
  fr_type temp = sat_sub(input, p->_State);
  p->_state += fr_mul(p->_pGain, temp);
  return sat_mul(temp, p->_dGain);
}
```

Listing 10.28 A differentiator using fractional arithmetic

PID Controllers

It is often a good idea to have a PID controller coded as its own entity, since it is used so often. This controller uses a different algorithm to limit the integrator state: rather than simply limiting the state to a fixed value, which tends to lead to overshoot, this controller limits the integrator depending on the amount of proportional control available. This has the effect of the system coming into a linear range of operation later, and with significantly less integrator windup.

```
typedef struct PID_struct
{
  fr_type _iState;       // integrator state
  fr_type _iMax, _iMin;  // integrator limits
  fr_type _iGain;        // integrator gain (always less than 1)
  int     _pGain;        // Proportional Gain (assumed to be > 1)
  fr_type _dState;       // differentiator state
  fr_type _dpGain;       // differentiator filter gain = 1 - pole
  int     _dGain;        // derivative gain * (1 - pole)
} PID_struct;

fr_type PID_update(PID_struct * p, fr_type input)
{
  fr_type retval, dTemp;  // return value, temporory differetial var
  fr_type pTerm = sat_mul(input, p->_pGain); // proportional

  // Find the integrator state, and limit it
  p->_iState = sat_add(p->_state, fr_mul(p->_iGain, input));
  retval = sat_add(p->_iState, pTerm);
  if (retval > p->_iMax)
  {
    p->_iState = sat_sub(p->_iMax, pTerm);
    retval = p->_iMax;
  }
  else if (retval < p->_min)
  {
    p->_iState = sat_sub(p->_iMin, pTerm);
    retval = p->_iMin;
  }

  // Calculate the differentiator state
  dTemp = sat_sub(input, p->_dState);
  p->_dState += fr_mul(p->_dpGain, dTemp);

  // Return the PID value
  return sat_add(retval, sat_mul(dTemp, p->_dGain));
}
```

Listing 10.29 A PID controller using fractional arithmetic

Example 10.1 Using the PID Controller

Figure 10.7 shows a block diagram of a motion control system. The plant is driven by a DAC (and a power amplifier that is not shown); its position is measured, then sampled by an ADC. The measured ADC value is compared to a target position and the resulting error signal is used to drive a PID controller.

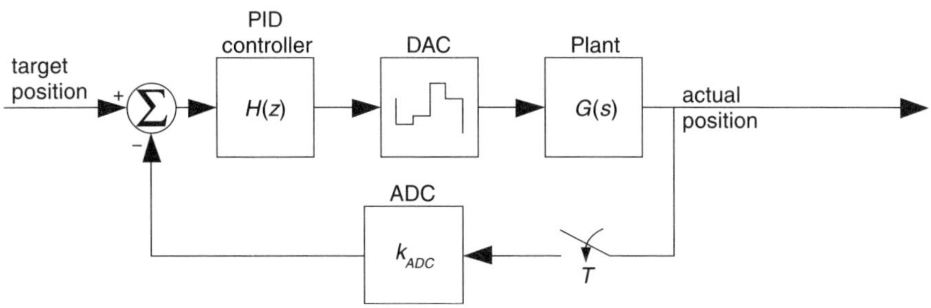

Figure 10.7 A motion control system

Assume that the system is set up so that the analog-to-digital converter is commanded to sample at a regular rate by hardware, and that the function `update_controller` gets called when the ADC completes its conversion. Assume that the function `get_adc_val` returns the raw ADC count as an integer, and that the function `put_dac_val` puts an integer to the DAC in raw form. Assume that the target position is accessed by calling the `get_target` function, with a return value of fr_type.

Write the `update_controller` function.

I will make the further assumption that the code from Listing 10.29 is available, and that the PID controller's definition structure has been correctly initialized. The code to actually update the controller is shown in Listing 10.30.

```
void update_controller(void)
{
  fr_type error;  // The loop error
  fr_type position, drive;

  position = div_to_fr(get_adc(), ADC_MAX);
  error = sat_sub(get_target(), position);

  drive = update_pid(&controlDef, error);
  put_dac(sat_mul(drive, DAC_MAX);
}
```

Listing 10.30 Using the PID controller

10.6 Conclusion

The activities shown in this chapter will be central to any controller coding effort. I have presented several approaches to implementing controllers in software, and have discussed most of the pitfalls that one can run into in implementing embedded control system code.

11

Afterword

In this book I have tried to present that cross section of the control systems field that will be of greatest benefit to the embedded systems engineer who is called upon to design control systems. This set of information isn't the whole of control theory, nor does it follow closely any academic program.

What this information *is*, however, is the part of control theory that I feel is the most pertinent and useful in day-to-day system design. I hope you find it as useful as I do.

11.1 Tools

You didn't think that I did all that math myself, did you? This is a technical book, with many equations and with examples of analysis or simulated results. Several pieces of software were used to generate the equations and the graphs.

Mathcad®, version 12, released 2004. Available from Mathsoft® (http://www.mathsoft.com).

Most of the equations that are derived symbolically in this book were done so—or at least checked for accuracy—using Mathcad. I try to use as many free tools as I can; Mathcad is an exception to this because it's worth the money. Mathcad provides a sort of free-form spreadsheet that lets you mix text, symbolic manipulations, and computations all on one page. With a bit of care you can perform the design computations for a control system in Mathcad in a format that can be printed out and used as a report.

Not only does it perform calculations, but it also allows you to do symbolic calculations, and it will carry units with every variable, allowing you do do automatic dimensional analysis as you go. These two latter features are what makes it a worthwhile buy for me.

Mathcad is the only product of its type that I have used personally, but if you are interested you should also look at Maple and Mathmatica.

Scilab, version 3.1. Free, available from http://www.scilab.org.

Similar to Matlab (*not* Mathcad) in philosophy but not compatible in the details (although it includes file conversion utilities). Scilab provides good number crunching ability, and comes with a very complete set of control system analysis tools. Includes native transfer function and state-space system representation types. Nearly all of the graphs in the manuscript for this book were calculated and generated in Scilab. When I am working I probably spend over 50% of my time in Scilab.

Scilab is an excellent environment for doing control system design; the structure of its built-in interpreted language is probably more suited for 'Matlab' sort of work than Matlab is. Its most serious drawback at this moment is the level of polish, and the documentation—it takes some digging to figure out how to make it do what you want, and there are some unfortunate inconsistencies in the way that function arguments are designed.

Alternatives to Scilab are *Octave*, a free application that is much more compatible with Matlab, and Matlab itself (*http://www.mathworks.com/*). At this writing, Scilab is much more refined than Octave, but it lacks the bells and whistles of Matlab. Matlab is the industry standard, but it doesn't come cheap. If you work in a well-funded corporate environment, then Matlab is a useful tool to have—otherwise get Scilab.

OpenOffice.org, aka *OOo*. Cleverly, this is the name of the website and the tools. A complete, free office suite. The manuscript for this book was entered in OOo Writer, and many of the figures were originally generated in OOo Draw.

gnu tools, cygwin version. See *http://www.gnu.org/* and *http://www.cygwin.com/*. All example code was compiled and tested using gnu C or C++ hosted in a cygwin shell on a PC running Windows.

11.2 Bibliography

This bibliography does not necessarily comprise the best available books on the subject, nor can I recommend one of these over any other in its field. The list below *is* the set of books that I found myself referring to as I wrote this book, and they are all at least minimally useful.

Adaptive Control, Karl J. Åström and Björn Vittenmark, Addison-Wesley 1995, second edition, ISBN 0-201-55866-1.

An advanced book, it has a good section on least-squares (ARMA) methods for system identification in Chapter 2, while Chapters 8 through 10 have some good information on not-quite-adaptive systems that are generally easier to apply in practice.

Signals and Systems, Alan V. Oppenheim, Alan S. Willsky with Ian T. Young, Prentice-Hall, 1983, ISBN 0-13-809731-3.

A classic text, to be seen on the bookshelf of every signal processing engineer of a certain age. Reviews partial fraction expansion in an appendix. Goes through the theory of signal processing in great detail; much of the signal processing theory in this book can be found, in unmutilated form, in *Signals and Systems*.

Engineering and Scientific Computing with Scilab, Claude Gomez, editor, Birkhäuser, 1999, ISBN 0-8176-4009-6, 3-7643-4009-6.

The closest thing there is to a comprehensive user's manual for Scilab; it covers Scilab functionality as well as techniques for solving engineering and scientific problems in a computational environment.

Understanding Digital Signal Processing, 2nd edition, Rick Lyons, Prentice Hall, 2004, ISBN 0-13-108989-7.

Just what it says it is. *Understanding Digital Signal Processing* is written for nearly the same audience that this book is, except that it covers signal processing where this one covers control systems. Where this book stops short on a DSP subject, *Understanding Digital Signal Processing* should be of help.

Design of Feedback Control Systems, Gene H. Hostetter, Clement J. Savant, Jr., Raymond T. Stefani, Holt, Reinhart and Winston, 1982, ISBN 0-03-057593-1.

A standard third-year college introduction to continuous-time control theory, it has a great deal more techniques for computing control solutions by hand, and a great deal less practical knowledge. This would be a good place to learn more about generating root locus plots. If you are willing to plow through the math, *Design of Feedback Control Systems*, or books like it, are a good adjunct to this book.

Feedback Control of Dynamic Systems, Gene F. Franklin, J. David Powell, Abbas Emami-Naeini, Prentice-Hall, 2002, ISBN 0-13-032393-4.

Like *Design of Feedback Control Systems*, this is a third-year college introduction. It also goes into great detail on root locus plots, as well as giving a fairly good outline of state-space design. Has a review of complex number theory in an appendix.

Digital Control Systems: Theory Hardware, Software, Constantine H. Houpis, Gary B. Lamont, McGraw-Hill, 1985, ISBN 0-07-030480-7.

Written when the potential of embedded control could be seen, but when only the most advanced systems used it. Written for fourth-year and graduate level university courses. Has rigorously correct math, while maintaining a good connection to practical reality.

Nonlinear Dynamical Systems, 2nd edition, Peter A. Cook, Prentice Hall, 1995, ISBN 0-13-625161-7.

A standard treatment of nonlinear control systems issues. Has a good expository style; intended for fourth-year or graduate students in control—but nonlinear systems theory requires some common sense along with the math, so portions are accessible to a wider audience.

The Art of Electronics, 2nd edition, Paul Horowitz, Winfield Hill, Cambridge University Press, 1989, ISBN 0-521-37095-7.

Just what the title says. If you only have one electronics text on your shelf, this should be it. Covers the design of electronic circuits from the basics to fairly advanced circuits, and does so in an open, understandable style.

About the Author

With formal training in analog circuits, communications systems and control systems (but not software), Tim Wescott probably has a typical resume for an embedded software engineer.

Mr. Wescott was unwillingly seduced into designing embedded systems during the course of developing the radio around which he wrote his master's thesis in electrical engineering (at Worcester Polytechnic Institute in Worcester, Massachusetts). In his search for a stable time base for the oscillators in his demodulator, his eyes fell on the crystal attached to the innocent little 8-bit microprocessor that had been intended to run the front panel. Several months of frustrating development later, the microprocessor was demodulating data at near-optimal levels, while only using 98% of the available processing power.

The early years of his career were spent in a futile rear-guard action, evaluating every design opportunity for its fit to an all-analog solution: each and every one ended up with processors doing the signal processing.

After giving up hope of using his analog circuit expertise in the design of analog circuits, Mr. Wescott took a job at FLIR Systems writing embedded software, much of it closing control loops. During his tenure at FLIR he noticed that not all of his embedded software colleagues shared his knowledge of control systems theory or application. He began giving lectures at local, then national, events. The oft-repeated question "Is there a book that covers this stuff?" became the genesis for this book.

Currently the owner of Wescott Design Services, Mr. Wescott has over fifteen years of experience designing and implementing embedded automatic control systems.

Index

G

H

I

L

M

N

O

ELSEVIER SCIENCE CD-ROM LICENSE AGREEMENT

PLEASE READ THE FOLLOWING AGREEMENT CAREFULLY BEFORE USING THIS CD-ROM PRODUCT. THIS CD-ROM PRODUCT IS LICENSED UNDER THE TERMS CONTAINED IN THIS CD-ROM LICENSE AGREEMENT ("Agreement"). BY USING THIS CD-ROM PRODUCT, YOU, AN INDIVIDUAL OR ENTITY INCLUDING EMPLOYEES, AGENTS AND REPRESENTATIVES ("You" or "Your"), ACKNOWLEDGE THAT YOU HAVE READ THIS AGREEMENT, THAT YOU UNDERSTAND IT, AND THAT YOU AGREE TO BE BOUND BY THE TERMS AND CONDITIONS OF THIS AGREEMENT. ELSEVIER SCIENCE INC. ("Elsevier Science") EXPRESSLY DOES NOT AGREE TO LICENSE THIS CD-ROM PRODUCT TO YOU UNLESS YOU ASSENT TO THIS AGREEMENT. IF YOU DO NOT AGREE WITH ANY OF THE FOLLOWING TERMS, YOU MAY, WITHIN THIRTY (30) DAYS AFTER YOUR RECEIPT OF THIS CD-ROM PRODUCT RETURN THE UNUSED CD-ROM PRODUCT AND ALL ACCOMPANYING DOCUMENTATION TO ELSEVIER SCIENCE FOR A FULL REFUND.

DEFINITIONS

As used in this Agreement, these terms shall have the following meanings:

"Proprietary Material" means the valuable and proprietary information content of this CD-ROM Product including all indexes and graphic materials and software used to access, index, search and retrieve the information content from this CD-ROM Product developed or licensed by Elsevier Science and/or its affiliates, suppliers and licensors.

"CD-ROM Product" means the copy of the Proprietary Material and any other material delivered on CD-ROM and any other human-readable or machine-readable materials enclosed with this Agreement, including without limitation documentation relating to the same.

OWNERSHIP

This CD-ROM Product has been supplied by and is proprietary to Elsevier Science and/or its affiliates, suppliers and licensors. The copyright in the CD-ROM Product belongs to Elsevier Science and/or its affiliates, suppliers and licensors and is protected by the national and state copyright, trademark, trade secret and other intellectual property laws of the United States and international treaty provisions, including without limitation the Universal Copyright Convention and the Berne Copyright Convention. You have no ownership rights in this CD-ROM Product. Except as expressly set forth herein, no part of this CD-ROM Product, including without limitation the Proprietary Material, may be modified, copied or distributed in hardcopy or machine-readable form without prior written consent from Elsevier Science. All rights not expressly granted to You herein are expressly reserved. Any other use of this CD-ROM Product by any person or entity is strictly prohibited and a violation of this Agreement.

SCOPE OF RIGHTS LICENSED (PERMITTED USES)

Elsevier Science is granting to You a limited, non-exclusive, non-transferable license to use this CD-ROM Product in accordance with the terms of this Agreement. You may use or provide access to this CD-ROM Product on a single computer or terminal physically located at Your premises and in a secure network or move this CD-ROM Product to and use it on another single computer or terminal at the same location for personal use only, but under no circumstances may You use or provide access to any part or parts of this CD-ROM Product on more than one computer or terminal simultaneously.

You shall not (a) copy, download, or otherwise reproduce the CD-ROM Product in any medium, including, without limitation, online transmissions, local area networks, wide area networks, intranets, extranets and the Internet, or in any way, in whole or in part, except that You may print or download limited portions of the Proprietary Material that are the results of discrete searches; (b) alter, modify, or adapt the CD-ROM Product, including but not limited to decompiling, disassembling, reverse engineering, or creating derivative works, without the prior written approval of Elsevier Science; (c) sell, license or otherwise distribute to third parties the CD-ROM Product or any part or parts thereof; or (d) alter, remove, obscure or obstruct the display of any copyright, trademark or other proprietary notice on or in the CD-ROM Product or on any printout or download of portions of the Proprietary Materials.

RESTRICTIONS ON TRANSFER

This License is personal to You, and neither Your rights hereunder nor the tangible embodiments of this CD-ROM Product, including without limitation the Proprietary Material, may be sold, assigned, transferred or sub-licensed to any other person, including without limitation by operation of law, without the prior written consent of Elsevier Science. Any purported sale, assignment, transfer or sublicense without the prior written consent of Elsevier Science will be void and will automatically terminate the License granted hereunder.

TERM

This Agreement will remain in effect until terminated pursuant to the terms of this Agreement. You may terminate this Agreement at any time by removing from Your system and destroying the CD-ROM Product. Unauthorized copying of the CD-ROM Product, including without limitation, the Proprietary Material and documentation, or otherwise failing to comply with the terms and conditions of this Agreement shall result in automatic termination of this license and will make available to Elsevier Science legal remedies. Upon termination of this Agreement, the license granted herein will terminate and You must immediately destroy the CD-ROM Product and accompanying documentation. All provisions relating to proprietary rights shall survive termination of this Agreement.

LIMITED WARRANTY AND LIMITATION OF LIABILITY

NEITHER ELSEVIER SCIENCE NOR ITS LICENSORS REPRESENT OR WARRANT THAT THE INFORMATION CONTAINED IN THE PROPRIETARY MATERIALS IS COMPLETE OR FREE FROM ERROR, AND NEITHER ASSUMES, AND BOTH EXPRESSLY DISCLAIM, ANY LIABILITY TO ANY PERSON FOR ANY LOSS OR DAMAGE CAUSED BY ERRORS OR OMISSIONS IN THE PROPRIETARY MATERIAL, WHETHER SUCH ERRORS OR OMISSIONS RESULT FROM NEGLIGENCE, ACCIDENT, OR ANY OTHER CAUSE. IN ADDITION, NEITHER ELSEVIER SCIENCE NOR ITS LICENSORS MAKE ANY REPRESENTATIONS OR WARRANTIES, EITHER EXPRESS OR IMPLIED, REGARDING THE PERFORMANCE OF YOUR NETWORK OR COMPUTER SYSTEM WHEN USED IN CONJUNCTION WITH THE CD-ROM PRODUCT.

If this CD-ROM Product is defective, Elsevier Science will replace it at no charge if the defective CD-ROM Product is returned to Elsevier Science within sixty (60) days (or the greatest period allowable by applicable law) from the date of shipment.

Elsevier Science warrants that the software embodied in this CD-ROM Product will perform in substantial compliance with the documentation supplied in this CD-ROM Product. If You report significant defect in performance in writing to Elsevier Science, and Elsevier Science is not able to correct same within sixty (60) days after its receipt of Your notification, You may return this CD-ROM Product, including all copies and documentation, to Elsevier Science and Elsevier Science will refund Your money.

YOU UNDERSTAND THAT, EXCEPT FOR THE 60-DAY LIMITED WARRANTY RECITED ABOVE, ELSEVIER SCIENCE, ITS AFFILIATES, LICENSORS, SUPPLIERS AND AGENTS, MAKE NO WARRANTIES, EXPRESSED OR IMPLIED, WITH RESPECT TO THE CD-ROM PRODUCT, INCLUDING, WITHOUT LIMITATION THE PROPRIETARY MATERIAL, AN SPECIFICALLY DISCLAIM ANY WARRANTY OF MERCHANTABILITY OR FITNESS FOR A PARTICULAR PURPOSE.

If the information provided on this CD-ROM contains medical or health sciences information, it is intended for professional use within the medical field. Information about medical treatment or drug dosages is intended strictly for professional use, and because of rapid advances in the medical sciences, independent verification f diagnosis and drug dosages should be made.

IN NO EVENT WILL ELSEVIER SCIENCE, ITS AFFILIATES, LICENSORS, SUPPLIERS OR AGENTS, BE LIABLE TO YOU FOR ANY DAMAGES, INCLUDING, WITHOUT LIMITATION, ANY LOST PROFITS, LOST SAVINGS OR OTHER INCIDENTAL OR CONSEQUENTIAL DAMAGES, ARISING OUT OF YOUR USE OR INABILITY TO USE THE CD-ROM PRODUCT REGARDLESS OF WHETHER SUCH DAMAGES ARE FORESEEABLE OR WHETHER SUCH DAMAGES ARE DEEMED TO RESULT FROM THE FAILURE OR INADEQUACY OF ANY EXCLUSIVE OR OTHER REMEDY.

U.S. GOVERNMENT RESTRICTED RIGHTS

The CD-ROM Product and documentation are provided with restricted rights. Use, duplication or disclosure by the U.S. Government is subject to restrictions as set forth in subparagraphs (a) through (d) of the Commercial Computer Restricted Rights clause at FAR 52.22719 or in subparagraph (c)(1)(ii) of the Rights in Technical Data and Computer Software clause at DFARS 252.2277013, or at 252.2117015, as applicable. Contractor/Manufacturer is Elsevier Science Inc., 655 Avenue of the Americas, New York, NY 10010-5107 USA.

GOVERNING LAW

This Agreement shall be governed by the laws of the State of New York, USA. In any dispute arising out of this Agreement, you and Elsevier Science each consent to the exclusive personal jurisdiction and venue in the state and federal courts within New York County, New York, USA.